Calculus Mysteries and Thrillers

© 1998 by

The Mathematical Association of America (Incorporated)

Library of Congress Catalog Card Number 98-85638

ISBN 0-88385-711-1

Printed in the United States of America

Current Printing (last digit):

10 9 8 7 6 5 4 3 2 1

Calculus Mysteries and Thrillers

R. Grant Woods
The University of Manitoba

Published and Distributed by
THE MATHEMATICAL ASSOCIATION OF AMERICA

Classroom Resource Materials is intended to provide supplementary classroom material for students—laboratory exercises, projects, historical information, textbooks with unusual approaches for presenting mathematical ideas, career information, etc.

101 Careers in Mathematics, edited by Andrew Sterrett
Calculus Mysteries and Thrillers, R. Grant Woods
Combinatorics: A Problem Oriented Approach, Daniel A. Marcus
Elementary Mathematical Models, Dan Kalman
Interdisciplinary Lively Application Projects, edited by Chris Arney
Laboratory Experiences in Group Theory, Ellen Maycock Parker
Learn from the Masters, Frank Swetz, John Fauvel, Otto Bekken, Bengt Johansson, and Victor Katz
Mathematical Modeling for the Environment, Charles Hadlock
A Primer of Abstract Mathematics, Robert B. Ash
Proofs Without Words, Roger B. Nelsen
A Radical Approach to Real Analysis, David M. Bressoud
She Does Math!, edited by Marla Parker

MAA Service Center
P. O. Box 91112
Washington, DC 20090-1112
1-800-331-1622   fax: 1-301-206-9789

To my calculus students, past, present, and future.

Contents

The Projects

The Solutions

The Purpose of This Book

This book consists of eleven calculus "research projects" embedded in short stories. Each project is based on some part of the curriculum of the "standard" two-semester calculus course on the differentiation and integration of functions of a single variable. The required "solution" to a problem usually consists of a report to a "client" (a hockey coach, a pirate captain, a fishing lodge owner) about how calculus can be used to solve the problem posed by the client. The recommendations in the report must be comprehensible to the client, who typically knows no calculus, and must also be supported by a detailed mathematical analysis, so that any "math expert" that the client might hire in the future can understand the reasoning leading to the recommendation.

The first goal of these projects is to develop skills in using calculus to analyze, and provide answers to, complex problems not presented in the language of mathematics. This of necessity involves having enough conceptual insight and technical ability to use calculus effectively in modeling situations. The second goal is to develop students' ability to write clear, complete accounts of how calculus has been used to solve complex problems. This requires facility at expository technical writing. Many students who use university level mathematics later in their careers will need modeling and technical writing skills more than technical mastery of concepts from more advanced areas of mathematics. "Writing across the curriculum" is in fashion nowadays, and this affords a way to implement it in a mathematics course.

The third goal is to provide raw material for instructors who wish to use "group projects" in their introductory calculus course as a means to encourage students to co-operate and talk together about mathematics. With perhaps two exceptions ("Calculus for Climatologists" and "The Case of the Cooling Cadaver"), these problems are too complex for one student to solve completely and write up in a two-week period. Much has been written about the importance of getting students to talk to each other about mathematics; these problems, used as group projects, provide a means to accomplish this.

At this point I must acknowledge a debt to a marvellous little book published by the MAA, entitled *Student Research Projects in Calculus*, by Cohen, Gaughan, et.al. (ISBN 0-88385-503-8) . It provides a detailed, carefully considered guide to the use of group "research projects" as a tool in teaching calculus. I first used the methods it describes to introduce group projects into a junior-level course in metric space theory taught to a small class of mathematics majors. Encouraged by the success of that experience, I next used it on two occasions when teaching a small honors class in introductory calculus. This led to the creation of seven of the eleven "short story" projects appearing in this book. Indeed, my first "short story" ("The Case of the Swiveling Spotlight") was inspired by a desire to elaborate on an essentially identical problem ("Calculus in the Courtroom") appearing in *Student Research Projects in Calculus*. I thank the MAA for giving me permission to crib from this earlier work.

Student Research Projects in Calculus contains an excellent account of how to organize and evaluate group projects, so I will not dwell on this subject here. Instead, I will urge those of you who wish to use these projects in a group setting to consult *Student Research Projects in Calculus* and follow their advice. My experience is that it works.

What distinguishes these eleven projects from those described in *Student Research Projects in Calculus* is that these are all detailed, sometimes lengthy short stories. My own experience, and apparently that of the authors of *Student Research Projects in Calculus*, is that students are often much more enthusiastic about "applied" problems of this sort than they are about "pure mathematics" problems, which are sometime viewed as boring and pointless. Furthermore, asking students to write a report (with its challenge to describe the solution of a mathematical problem in non-mathematical terms) is more natural in this setting. These stories also contain "local color" (and some attempts at humor) that are extraneous to the mathematics. Again, my experience is that many students appreciate the effort to "humanize" what is perceived by many to be a rather cold and austere subject. I apologize to any hockey fans, ufologists, pirate captains, lawyers and others whom I may have inadvertently offended in the process.

The book closes with a collection of model solutions. Students' solutions are seldom as complete as these, but these can serve as a illustration of what could be done. I suggest that you read the model solution before assigning a problem to see in detail what calculus requirements, and hidden difficulties, are lurking in the problem. Knowing this in advance may help you to avoid problems later.

Finally, the Canadian overtones of several of the projects are explained by the fact that the author is a Canadian.

An Overview of the Projects

The order in which the projects appear in this book follows (roughly) the order in which the concepts they utilize are introduced into the typical single variable calculus course. Several projects use a number of different ideas, so this order is of necessity only approximate.

I have assigned seven of the eleven projects to my calculus classes (all but "Calculus for Climatologists," "Designing Dipsticks," "Sunken Treasure," and "The Case of the Alien Agent"). The University of Manitoba is similar to a big American state university, and there is a very large range of ability and preparation among our incoming calculus students. The students in the calculus courses in which I used these projects were probably a random selection from the top 25% of our freshman calculus students. This experience has convinced me that these seven projects are feasible for reasonably bright freshmen working in groups. It has also led to my classification of each as easy, moderate, or difficult. My classifications of the remaining four projects are informed guesses.

Obviously these projects are not particularly accurate descriptions of reality, or of situations in which calculus is actually used. In my experience this does not interfere with their pedagogical value—students accept the projects in the spirit in which they were written. (Occasionally I have included a disclaimer; see "An Income Policy for Mediocria" and "The Case of the Cooling Cadaver.") In fact, one group appended to their solution of "The Case of the

Cooling Cadaver" a copy of an article from a medical journal in which the applicability of Newton's Law of Cooling to the recently deceased was refuted and another model was put forth in its place. They then proceeded to use this other model to show that based on the given data, the victim had been murdered several hours before he was last seen alive. Any mathematics project that provokes a response like that should be counted a success.

Feel free to modify projects. Dropping the requirement to use Newton's Method from "The Case of the Swiveling Spotlight" will make it more accessible, as will omitting the final problem in "Designing Dipsticks" in which the dipsticks are accidentally interchanged. In the opposite direction, I toyed with the idea of the murderer in "The Case of the Cooling Cadaver" turning up the thermostat in the study upon leaving so that the temperature of the study increased linearly as time passed. Newton's Law of Cooling then gives a more involved (but easily solved) linear first order differential equation, and the problem becomes more difficult. You will undoubtedly have your own ideas.

It is a challenge to create applied problems that assume no knowledge of any specific applied field (such as physics or economics). As most students studying calculus know some analytic and Euclidean geometry, many of these problems have a geometric foundation. Instructors whose students have homogeneous backgrounds could easily create problems drawing on this common background. If, for example, one's class consisted entirely of engineering students who knew some classical mechanics, whole vistas of interesting plot possibilities immediately open up.

Two other themes run through these problems. In many of them an important first step is to impose a system of co-ordinate axes on the underlying geometry, and students may need to be told about the importance of this. Second, deciding on the appropriate domain of definition of some of the functions that arise is also very important. One of the virtues of these problems is that they put students into the habit of thinking about domains.

Finally, several problems finish by asking students to perform a numerical calculation for a specific case. This helps test whether students really understand what they have done.

Detailed Mathematical Requirements

Project Title: The Case of the Parabolic Pool Table
Difficulty Level: Easy
Mathematical Requirements: Students must know how to introduce Cartesian co-ordinate axes into a geometric situation in a well-chosen way. They must know the form of the equation of a parabola. They must know about normal lines, how to find the slope of a line through two given points, and how to differentiate polynomials.

They must realize the necessity of verifying that the X-co-ordinate of their "solution point" lies within an appropriate set of values, and that negative numbers do not have real square roots.

Project Title: Calculus for Climatologists
Difficuly Level: Easy
Mathematical Requirements: This is the easiest of the projects; once one sees how to do it, a detailed exposition of the solution can be very brief. Its attraction is that the result is both mildly counterintuitive (at first sight) and quickly achieved.

The only mathematical requirement is the Intermediate Value Theorem for continuous functions on closed intervals. It requires ingenuity, and one hopes that it will cause students to think carefully about what the Intermediate Value Theorem actually says.

This project has not been assigned to a class.

Project Title: The Case of the Swiveling Spotlight
Difficulty Level: Difficult
Mathematical Requirements: I've classified this as difficult because it would probably be assigned at an early point in the course, and it draws from several different knowledge areas. It requires all the ideas described in the first paragraph of the Mathematical Requirements for " The Case of the Parabolic Pool Table" (see above), plus the following: an ability to sketch the graph of a polynomial (specifically an increasing cubic polynomial), and an understanding of Newton's Method. This latter could be dropped and replaced by the use of a scientific calculator or computer algebra system to approximate roots of cubic equations.

Project Title: Finding the Salami Curve
Difficulty Level: Moderate
Mathematical Requirements: This is another problem in which the first challenge is setting up and using an appropriate co-ordinate system. The problem requires students to optimize a function on a closed interval. The function is the difference of arctans, so students either need to know how to differentiate $\arctan x$ or else know how to get the same result by implicit differentiation and the differentiation of $\tan x$. They need to know the basic rules of differentiation, including the chain rule and the quotient rule. Classification of critical points using the first derivative test is also used. Mastery of curve-sketching techniques will be helpful.

The graphing should be straightforward, and those within easy telephone reach of a National Hockey League franchise can find the numerical data needed to do the final question.

Project Title: Saving Lunar Station Alpha
Difficulty Level: Moderate
Mathematical Requirements: This problem involves both related rates and minimization of a function defined on an interval. Students must know how to use the Chain Rule, how to differentiate polynomials and power functions, and how to go about classifying a critical point of a function. Some command of Euclidean geometry and trigonometry is also required. The latter part of the solution involves some calculator work.

Project Title: The Case of the Cooling Cadaver
Difficulty Level: Easy
Mathematical Requirements: This problem is an application of Newton's Law of Cooling. Students must therefore know how to differentiate exponential functions. The situation reduces to solving three linear equations in three unknowns, and because exponential functions are involved, this requires some facility with the algebra of exponential functions. Students may have difficulty deciding when to start counting when they define the independent variable "time." Some calculator work is needed at the end.

Project Title: An Income Policy for Mediocria
Difficulty Level: Difficult
Mathematical Requirements: The objective of this problem is to give students practice in "setting up" a definite integral as a limit of a sum in order to model an economic situation (rather than a situation drawn from geometry or physical science). The economic and social arrangements in Mediocria are undoubtedly bizarre, but the problem seems to work.

To do this problem students need to have seen examples of how to "set up" a definite integral as a limit of a sum in order to model some phenomonon. They need to know the Fundamental Theorem of Calculus and how to integrate polynomial functions. Integration by parts is also used. Students may need more than the usual ability at interpreting the meaning of mathematical results in terms of the situation being modelled. The fact that $\int_a^b f(x)\,dx = \int_a^c f(x)\,dx + \int_c^b f(x)\,dx$ is also used.

Students who have studied continuous probability distribution functions in a statistics course will find this problem significantly easier than those who have not.

Project Title: Designing Dipsticks
Difficulty Level: Moderate
Mathematical Requirements: This problem requires students to calculate volumes of solids of revolution of a portion of a sphere, and areas of portions of a circle (both by using integration and by using Euclidean geometry and trigonometry). The former involves integrating a degree 2 polynomial, while the latter involves integrating $\sin^2 x$ and using the arccos function. Some facility at using trig and inverse trig functions is required. To calibrate and physically produce two dipsticks requires considerable calculator or computer algebra software work; this project involves more of this work than any other project.

The final question requires approximation of the roots of a cubic equation. This project has not been assigned to a class.

Project Title: The Case of the Gilded Goose-egg
Difficulty Level: Difficult
Mathematical Requirements: Students must compute the surface area of a strip of an ellipsoid of revolution, and then decide which of a set of such strips of equal thickness has the largest surface area. As we have set up the problem, this latter requires the use of the "$d/dx\left(\int_a^x f(t)\,dt\right) = f(x)$" version of the Fundamental Theorem of Calculus for a continuous function f. So, students must know how to compute the surface area of a solid of revolution, must know this version of the Fundamental Theorem of Calculus, and must know how to optimize a differentiable function on a closed interval; this latter calculation involves (in my model solution) use of the Intermediate Value Theorem. A complication is introduced by the fact that permissible solutions must occur at integer fractions of the length of the original ellipse.

The first part of this problem (showing that if a sphere is sliced into slices of equal thickness then each slice has the same "outer" surface area) is a standard exercise found in many calculus texts. Here it serves as a "warm-up" for the main event.

Project Title: Sunken Treasure
Difficulty Level: Difficult
Mathematical Requirements: This problem requires students to calculate the arc length of a portion of a parabola, and to find where a tangent to the parabola from a given point "outside" the parabola meets the parabola. Students must know how to use a definite integral to calculate arc length. In this particular situation this involves using a trig substitution and evaluating the ubiquitous integral $\int \sec^3 x\,dx$.

Setting up an appropriate co-ordinate system is an important initial step. The problem splits into two cases depending on the relative sizes of the distance parameters; recognizing this, and distinguishing between the two cases, is an important part of the problem. Students have to cope with long, messy mathematical expressions. There is a numerical calculation at the end to test if the students understand the interpretation of their results.

This is probably the most difficult of the projects. It has not been assigned to students.

Project Title: The Case of the Alien Agent

Difficulty Level: Difficult

Mathematical Requirements: This problem involves calculation of the volume of a solid of revolution by rotating about the Y-axis rectangular elements whose long dimension is neither horizontal nor vertical. To calculate the size of the resulting element of volume one needs to derive and then use the formula for the surface area of a portion of a right circular cone. The resulting integral involves some trigonometric manipulation, and evaluating it involves integration by parts and integration of $\sin x$ (or $\cos x$).

Again, setting up an appropriate co-ordinate system is an important initial step. Another important feature of this problem is the use of simplifying approximations based on the fact that one of the parameters is many orders of magnitude less than another. At the end a Maclaurin series is used to help approximate a magnitude; if infinite series is not part of your syllabus, some other method could be used to achieve a lower bound on the magnitude of $(6/\alpha^3)(\alpha - \sin \alpha)$. This project has not been assigned to students.

Problems

1

The Case of the Parabolic Pool Table

A hush descended over the classroom as Inspector McGee strode to the podium. After all, he had served on the Fraud Squad for over 20 years, and the talks that he gave to each year's crop of police recruits were the stuff of legend. No one else had the fund of stories about past experiences that he did. Each year he amazed the rookies by telling them of yet another bizarre technique that he had used to unmask the schemes of the city's con artists. It seemed that there was no branch of knowledge that he hadn't exploited at some point in his career. But this year, the rumor went, he was going to outdo himself; somehow he had actually used calculus to convict a criminal! It seemed hard to believe. The rookies strained forward in anticipation.

"First of all, I'd like to thank the Chief for giving me a chance to talk to you young recruits," McGee began. "My goal today is to convince you that you must always be ready to use experts to help you get to the bottom of a fraud scheme. In fact, in one of my most interesting cases I called on the smarts of a bunch of young people about the same age as you. Let me tell you about it.

"A few years ago a new bar opened in the Little Bohemia district—maybe you remember it. Upstairs it featured the usual assortment of rock bands, but the real attraction was in the basement. They called it Luigi's Lizard Room, and judging from the sleazy characters that hung out there, it was well-named. It was full of pool tables—beautiful, big tables with lots of room between them and well-enough lit so that you could see what you were doing. They even provided a little oxygen with the smoke. The place was always packed, and you often had to wait a half hour or more before you could get onto a table. They had a real neat gimmick—besides the usual rectangular tables, they had a bunch of custom-made tables in weird shapes; equilateral triangles, circles, and even ellipses. Banking shots off cushions was real complicated on those tables, and high school teachers all over the city reported a sudden upsurge of interest in analytic geometry.

"Anyway, one day we got an anonymous call from a young guy who sounded really irritated. The place was run by a pool shark (named Luigi, of course) and according to our informant Luigi was operating a little racket. We decided to go down and see for ourselves.

"When we got there there was a huge crowd gathered around two tables that were side by side. They looked similar to each other, but different from any pool table that I had seen before. Three sides were sides of a rectangle, but the fourth side was rounded outwards and looked something like this." McGee paused and drew a rough sketch on the blackboard. "The round part sort of formed a corner where it met the straight part—it wasn't really smooth. There were only five pockets on the table, in the places where I've drawn the little circles. Four were at corners, and the fifth was at the top of the round part.

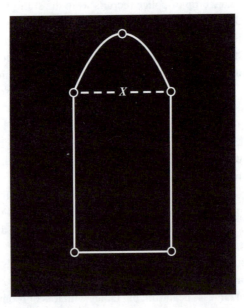

A parabolic pool table with pockets at corners and vertex.

"Anyway, Luigi was having a little bet with a customer. He used one table, and the sucker—sorry, customer—was using the other. Each of them placed a cue ball at the x in the middle of the missing side of the rectangle, and had to shoot the ball against the curved part of the cushion. The object was to get your ball to bounce back directly over the x where it had started from. You couldn't hit the cushion more than once. If you were successful, the other guy had to pay you five bucks. If you weren't, you didn't collect.

"We watched for about half an hour, and Luigi was getting rich fast. Some young stud would figure that he was sharp enough to take Luigi, but he'd never be able to get his ball back across the x. Luigi would make about half his shots, so eventually he would clean up. After a while the sucker would get suspicious and quit, but next evening there would be new victims. We figured there was something wrong, but we couldn't see what it was."

"Excuse me, sir," interrupted one of the recruits. "From your picture it looks as if each table was symmetric, so couldn't the sucker just shoot straight for the top of the curved part and have it bounce straight back over the x? That wouldn't be hard for a good player."

"You weren't listening carefully enough, son," sighed McGee. "There was a pocket up there. Balls don't bounce off pockets—they go in." The recruit shrank into his chair.

"I went home and thought about it. There had to be something wrong with the shape of the sucker's table, but I had no way to figure what it was. Over dinner next day I mentioned the situation to my kids, and warned them against going there. And that's when I got my big break. My son was an undergrad at the university back then, and he mentioned that when three of his friends who were math majors couldn't find summer jobs, they had formed their own consulting firm. So I looked up **Math Iz Us** and gave them a call the next day.

"You couldn't hope to meet three nicer people than Heather, Sasha, and Li. After I outlined the problem to them they paid a visit to the Lizard Room and watched Luigi at work. Two days later they called me back. They had made a key observation at the Lizard Room—attached to the sides of the two tables were small manufacturer's plates saying 'parabolic model.' They also listed the manufacturer's phone number. So Heather had phoned the manufacturer and confirmed that the curved portion of each table was indeed in the shape of a parabola. She tried to get more information about the precise shape, but the guy on the other end suddenly clammed up. We still have our suspicions about that outfit.

"Next day we sat down with **Math Iz Us** and reviewed the situation. They told us that they thought they had figured out what might be happening, but they needed just a little bit more information about the two tables. Before we went any further, we drew up a formal contract with them. Those kids made a nice bundle of money before the case was over!

"The next morning, posing as cleaning staff, they gained access to the Lizard Room and made some measurements of the two tables being used. Their rectangular portions had exactly the same dimensions, but the parabolic parts weren't quite the same—it bulged out farther on Luigi's table than it did on the table that the sucker used. For each of the tables they took precise measurements of the lengths of the two straight parallel sides, the width of the third side, and the perpendicular distance from the x to the top of the parabolic part. They were pretty sure that they only needed to know two of those three dimensions, but they wanted to be on the safe side. They didn't want to have to go back again!

"When they called us three days later they had great news. By using some analytic geometry and a bit of calculus they had **proved** that no matter where on the parabolic part the sucker bounced the ball, it would never bounce back over the x where it had started from! On Luigi's table, however, there were two places where, if the ball hit one, it would bounce straight back over the x. Not only that, there were faint ink marks at precisely those places on Luigi's table, so Luigi knew where to shoot! They were inconspicuous enough that you wouldn't notice them if you weren't looking for them. Luigi apparently purposely missed some of his shots so it wouldn't be too obvious a set-up, but when he wanted to score, he knew where to shoot.

"So, we arrested Luigi and charged him with fraud, but at the preliminary hearing we got some bad news. The case was being heard by Judge Vance Frito. Now Frito's dad was a math prof, and Frito is a chip off the old block. We just knew that he would want a complete analysis of the general situation! So we got Math Iz Us to write us a detailed report analyzing what the relative dimensions of a pool table of this sort had to be in order for there to be a place on the curved cushion where a ball could be bounced directly back over the x. They assigned letters to represent their three measurements, and then they showed that there was such a place if and only if the ratio of two of the three measurements that they took was less than a certain fixed number. That was the central conclusion of their report. After that all they had to do was submit the evidence about the measurements of the tables in the Lizard Room to show that there was such a place for Luigi's table, but not for the sucker's. We won the case, and now the Lizard Room is a ping-pong emporium.

"Later I asked Math Iz Us if they had found the problem to be difficult. They told me that setting it up was a lot harder than the actual calculations, and that the amount of calculus involved was actually rather small. Introducing co-ordinate axes in the right way was really the key. And of course, knowing the laws governing how balls bounce off cushions was important. Sasha said that the basic fact came to him when he was fooling around with one of the circular tables at the Lizard Room, and noticed that if he put his ball at the center of the table and bounced it off any cushion whatsoever, it would come right back over the center of the circle. Of course, none of those guys in the Lizard Room ever put any fancy spin on the ball, so the straightforward bouncing laws applied.

"Apparently the pressure was really on them when they realized that writing a clear, complete brief for Judge Frito was going to take them longer than they had anticipated. Not only is Frito a math freak, he's a real stickler for clear, correct English. So the brief had to be just right, and it took a couple of revisions to get it that way. And of course, it had to be submitted by the day of the trial. But they did a good job, and Judge Frito commended them for the quality of their work.

"So remember, recruits, a good fraud detective has to be willing to get help from anyone and everyone."

Prepare the brief that Math Iz Us submitted to Judge Frito. Since you are explaining the whole situation to the judge, make sure that your account is complete and self-contained. Frito is a math fan, so make sure you include in detail the mathematical reasoning that led you to your conclusion. The most important part of your report is the result about the ratio of two of your measurements (see above). You can make up some plausible data about the dimensions of Luigi's table and the sucker's table and then finish your report. Use diagrams, and describe precisely where on Luigi's table he should be shooting.

Make sure that your brief is submitted by the date of the trial on [assignment due date].

2

Calculus for Climatologists

Jason and Anne had been friends since grade school, so each of them was delighted when they learned that the other had been accepted into Megastudent University. Mega U. was situated in a small city several hundred miles from their home town, and although it had a reputation for academic excellence, some of their older friends had returned from their freshman year there with scary tales about how cold and impersonal the large campus could be for shy students. So that summer, as they prepared to leave their homes and friends, Jason and Anne made a pact that they would keep in touch and spend time together on campus.

This proved to be something of a challenge. Anne enrolled in a program consisting mostly of physical science and mathematics courses, while Jason was primarily interested in the social sciences. This meant that they seldom saw each other in class, and their heavy workloads made it difficult to find a lot of spare time outside of class. However, two circumstances kept them in touch; they were enrolled in the same course in introductory geography, and every Friday afternoon at 3:30 PM they would meet in Thirsty's, the campus hangout, to unwind from the week and catch up on news. They gossiped about their friends, complained about their professors, and talked about the ideas they were encountering in their courses. One favorite topic of conversation was their different outlooks on the power of science, and scientific modes of analysis, to explain the world around them.

"Physical science and mathematics are OK for explaining simple things like how the planets move and when the next eclipse will be," said Jason one day, when he had finished his drink and was starting to feel argumentative. "But they'll never be much use for explaining things like politics, and poverty, and why people act the way they do, and everybody knows that people are more important than the dead matter that you science types are always talking about. To understand people you need social science. Even literature is better than math for that."

Anne always confined herself to gingerale. She thought that it made it easier for her to stay calm and rational when Jason got into one of his belligerent moods. Jason was a lot of fun, but he liked arguing too much.

"But people do use math and science to help understand people," she replied. "Look at how they use chemistry to understand hormones, and the way the brain works, and how all that affects our behavior. And pollsters use statistics to predict how people will vote, and what they will buy. And my prof was telling us that even economists use mathematics to predict whether we'll have a recession...."

"Yeah, sure," retorted Jason. "Everybody knows that if you have 10 economists, you'll have 20 different theories about what's happening...they can't even agree among themselves! And those neuroscientist types just want to remove all the mystery and poetry from human existence. I think that...."

"Is this a private fight, or can anyone play?" Jason and Anne looked up to see Len standing next to their table. Len was in their geography class, but he was in his final undergraduate year and was supposed to be graduating in May. The geography course was a "filler" to complete his required credits.

"Sit down," said Jason. "Anne was claiming that physics and math can explain everything, and I was setting her straight on how it really is, and...."

"I was not!" exclaimed Anne. "You started this silly conversation, not me, and I never said anything about...."

"Children, children," said Len. "Jason, do you know any calculus?"

"A little bit," said Jason. "My dad made me take it in Grade 12."

"Well, maybe you two would like to make a little wager. Now, this geography course we're taking—is it physical science or is it about people?"

"Some of each, I'd say," said Anne. "The stuff we're taking right now about weather and climate seems more like physical science, though. It's really interesting."

"But it proves my point," said Jason. "I remember reading that whether or not a butterfly flaps its wings in India in July might determine how big a blizzard they would have in Denver in December. And there's no way the scientists can know all about what the butterflies are doing in India, so they can't predict the weather—at least, not a long way ahead."

"Actually, there's a mathematical theory—called chaos theory—behind what you said," replied Len. "But Jason, would you believe that mathematics all by itself, with just a very tiny bit of physics, can tell you something unexpected about weather and climate?"

"What's that?" said Anne, suddenly very interested.

"Right now there are two antipodal points on the equator that are at exactly the same temperature. Did you know that?"

"Remind me what antipodal points are," said Jason.

"Don't you remember?" asked Anne. "They're points on the same great circle that are 180° apart, like the North Pole and the South Pole on a circle of longitude, or like 90° East and 90° West on the equator. They're two points that are as far apart as they can be."

"You can't tell me that you know for sure that 90° East and 90° West on the equator are at exactly the same temperature right now!" exclaimed Jason.

"That's not what I said," replied Len. "What I said was that there would be **some** pair of antipodal points on the equator at the same temperature. I didn't say which pair it would be."

"Well, after all, it's hot all along the equator, so maybe it might work..." mused Jason.

"It's not that simple" retorted Len. "There's two antipodal points at the same temperature on any great circle through the North and South Poles, too... and they don't have to be the North and South Poles, either."

"That's weird!" said Anne. "I wonder what it is about great circles that...."

"Maybe less than you think," said Len. "Anyway, here's the wager, if you want to take it. The first person who produces a convincing argument that what I said is true gets the free beverage of their choosing from me here next Friday. If neither of you figures it out, each of you buys me something to drink. How does that sound to you?"

"Suppose I produce a convincing argument that you're full of baloney?" asked Jason.

"That's fine. You'll get your drink," replied Len. "After all, maybe I'm lying to you! Oh, yes, one thing—one little tiny piece of physics. Do you agree that temperature is a continuous function? You know the technical definition, but intuitively it means that nearby points will be close to the same temperature. Do you agree?"

"Seems reasonable to me," said Jason. "But what does that have to do with anything?"

"Think about what you know about continuous functions," replied Len.

"But in our calculus course we just talk about continuous functions defined on a line," said Anne. "The Earth is a two-dimensional surface, and we don't study that kind of calculus until next year."

"Yes, but circles are one-dimensional," replied Len. "Now go away and be clever."

Anne tried to use mathematics to show that Len was right, and Jason tried to think of a situation that would illustrate that Len was wrong. One of them was successful. Which was it? Present a **convincing and detailed** mathematical argument that either proves Len's assertion or else gives an example illustrating that it is wrong.

After the successful person's presentation was made to Len on the following Friday, the other person made an astute observation about the Tropic of Cancer that went beyond Len's assertion. What do you think it was? Explain in detail.

Your presentation could be in the form of a description of the meeting that the three of them had the following Friday at Thirsty's.

3

The Case of the Swiveling Spotlight

Hi. My name is Friday—Joe Friday. I'm a calculus student. I want to tell you a little story—a story about parabolic boulevards, focal fountains, and a car speeding through the night. I like to call it ... the Case of the Swiveling Spotlight.

It all started one bleak Monday evening in early November. I'd gotten home from classes, made some sandwiches, and settled down with a tall glass of something cold when the phone rang. It was my older sister Della. Della's a smart woman—she graduated from Law School a few years ago and now she's working for a law firm in Lac du Portage, a small town southeast of here. As soon as she started talking I could tell that she was worried.

"Joey, I've got a problem. A real problem. I've been hired to defend a local citizen on a charge of speeding. He's already had a few traffic citations, and if he's convicted this time he's sure to lose his license. Our only hope is to convince the judge that the police don't have enough evidence. But the case comes to trial in 17 days, and I need to have a written brief to present to the judge by that time. When I described the case to my secretary Perry, he told me that he thought that it might involve calculus. You're the only person I know whose calculus is up to date. Joey, you've got to help me!"

"Sure, sure," I replied, "just give me the facts." And she did.

"Late one night a police patrol car was cruising along the long curved boulevard in the shape of a perfect parabola, which passes through a large rectangular park in downtown Lac du Portage. Just as the cruiser passed through the vertex of the parabola Sergeant Preston gazed admiringly at the beautiful fountain that marks the focus of the parabola; it was located directly north of him to his left. At that very moment his partner Sergeant Renfrew noticed through the trees a suspicious vehicle—my client's vehicle, as it turned out— moving at a constant speed in an easterly direction along the street that bounds the park along its southern edge. Immediately turning the cruiser car's infrared spotlight onto the speeding car, Renfrew was at once able to determine the vehicle's speed, and how far east of him it was, by using his watch and the known distance between streetlights."

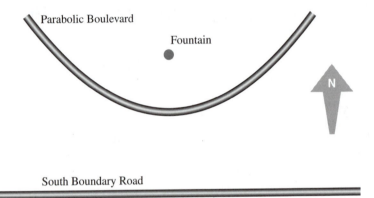

Park du Portage.

"And you're quarrelling with the accuracy of their determination of the vehicle's speed and its distance east of the cruiser?" I interrupted.

"No," Della replied. "We're willing to accept their figures, but we want to show that if their figure for the speed was correct, then they couldn't have been measuring the speed of my client's vehicle for as long a period of time as the law requires. Keep listening." So I shut up and did.

"The cruiser drove along the boulevard, continually adjusting its own speed so that it always remained the same distance west of my client's car." Della continued. "As Preston drove, Renfrew kept his infrared spotlight trained on the car. He was able to observe my client's actions for a while, but several seconds before his spotlight would have been pointing directly sideways, he found that it was stuck and would not pivot any further. Nevertheless, the officers felt that by this time they had enough information to proceed, and they arrested my client for speeding.

"The next morning Preston returned to the scene and carefully measured the distance from the vertex of the parabolic boulevard to the nearest point on the south boundary road. I sent Perry over there to measure that distance himself, and he got the same value as Preston did. So, everyone knows that distance and there's no dispute about its value.

"As you know from your own dubious experiences, Joey, the law requires at least 20 seconds of uninterrupted observation of the speeding car under these circumstances. We don't know how long Renfrew had his spotlight trained on my client's car, but we know that he quit before he had gone far enough for his spotlight to be pointing directly sideways from the patrol car. We were hoping that since we knew how fast the car was supposed to be going, we could show that the spotlight couldn't have been on it for as long as 20 seconds. Perry suggested that we should draw a picture of the park and the roads and stuff on some co-ordinate axes, but my geometry is rusty, and when Perry suggested using calculus, I knew I was beaten. I took introductory calculus ten years ago, but I've forgotten most of it. I sure hope you can help me."

It was an ugly problem, but someone had to solve it. And I knew just the bunch to do it—my Study Group from last semester's calculus class!

One hour and two phone calls later the Group was assembled around my kitchen table. There were three of us—Jason Wednesday, Nicole Thursday, and me. We had solved a couple of challenging problems before, but this one looked different. And we had another challenge—Della had forgotten to give me the numerical values for the three pieces of data whose values she knew. "Oh, well," I thought, "we can just use letters to represent them now and plug in their values later."

"The first thing we need to do," said Nicole after I had explained the situation to them "is to draw an accurate picture of the roads and the park boundaries, and show where the two cars are. Maybe if we draw it to scale on co-ordinate axes we can write some equations that will help describe the situation."

While Nicole did that, I thought about what it meant for the spotlight to be pointing directly sideways from the cruiser car. The phrase "normal line to a curve" kept rattling through my brain like the catchy chorus to a bad band tune. I decided to go to my calculus text and check it out.

After what seemed like several days (but it might have been only half an hour) Nicole sighed and stretched. "Well, boys, I've got good news and bad news," she said. "The good news is that if I knew the equation of that parabolic boulevard I could use some geometry to get an equation for the time that elapsed from when the spotlight was first turned on to the the time when the spotlight would have been pointing directly sideways had it continued to pivot. And clearly that time must be longer than the time that the spotlight was actually trained on the car."

"So if we could show that that time was less than 20 seconds . . . " I began. "That's right," said Nicole, "and by the way, thanks for that hint about normal lines. It was really helpful."

"What's the bad news?" inquired Jason.

"The bad news is that we don't have enough information to write down the equation of the parabolic boulevard," replied Nicole. "Until we do we're stuck."

"Wait a minute," said Jason. "Joe, didn't you say that there was a fountain at the focus of the parabola? Maybe if we knew how far it was from the vertex to that focus, Nicole would have the information she needs."

"That's right!" I exclaimed. "I can call Della and get her to have Perry measure that distance. Nicole, in the meantime why don't you just assign some letter to represent that distance, assume we know its value, and work from there?"

"I could if I knew how to use your information to work out the equation of the parabola," said Nicole sadly.

Silently cursing myself for not listening more carefully to my high school math teacher's lessons on parabolas, I dug back into my calculus text to extract the needed information. Luckily it had an appendix on conic sections. When I was done I told Nicole what she needed to know, and she completed her task.

Now it was Jason's turn. He's our algebra specialist. He looked at Nicole's equation, manipulated it a bit, scribbled some calculations, and did a quick sketch. Then he turned to us. "You can see that it's in the form of a third-degree polynomial," he said. "The time that we want is a root of it."

"Big deal," I replied, "cubic equations can have as many as three different roots. How do we know which one is the right one?"

"Look at the coefficients," said Jason, "and by the way, I've used single letters to represent them to keep things from getting too messy. I used calculus techniques to graph that polynomial—see, here's the picture. You can see immediately that the polynomial has only one real root and it is positive. Actually, if the judge is as much of a stickler for rigorous arguments as your sister says he is, I could even prove it with the Intermediate Value Theorem."

"Maybe you should," I replied. "We don't want Della to lose on a technicality. And you better be able to convince the judge that your graph is correct."

"But we still have problems," said Nicole. "It's hard to solve a cubic equation exactly. And unless we do, how will we know whether its root is less than 20 ?"

Suddenly I remembered a conversation with an acquaintance of mine who's in second year computer science. He had told me about something called Newton's Method for approximating roots to equations. Now I never talk to computer scientists if I don't have to, and in this case it wasn't necessary; I grabbed my trusty calculus text again and read all that I needed to know there. (By now I was glad that my prof had gotten me into the habit of looking things up in the text if they hadn't been covered in class.)

"Listen up," I told the others. "Newton's method is an iterative procedure. If I put an initial guess for the root into the procedure, it spits out a second approximation. Hmm, let's start with...." I scribbled for a while. Then I exclaimed, "Look; you can see from Jason's graph that the second approximation is bigger than the actual root!"

"That's great!" said Nicole. "And we'll be able to work out that second approximation exactly when we have the numerical data from your sister. If it's less than 20, then Della will win her case!"

"And even if it's not less than 20, if we apply Newton's method a few more times we might get a better approximation that will do the job," said Jason. "I wonder if we can convince the judge of that."

"And perhaps we need a better explanation of why that second approximation is bigger than the real root," mused Nicole. "After all, they say that justice is blind, so maybe the judge won't be convinced by looking at a picture."

"He ought to be," I said, "but maybe I should include a rigorous proof in our brief. Let's see, if I use the Mean Value theorem here...."

It went quickly from there. Of course, it took longer than we expected to write the brief (it always does), but we faxed it to Della in time for her to submit it to the judge on the day before the trial. It's funny—when we sent Della the brief we had no idea whether or not it would prove her contention that her client's car had been under observation for less than 20 seconds, because we still didn't have the numerical values for the distances and purported car speed that we needed. Della had to fill those in and do the arithmetic herself. But a few days after the trial she called me to say that her client had been acquitted and that our brief was the deciding factor! So I guess the arithmetic worked out OK. And in his decision, the judge praised our Calculus Study Group and said we were a fine bunch of mathematicians!

Write the brief that Joe's Problem Solving Group faxed to Della. Since the judge is an old grouch who is a stickler for deadlines, make sure that it is submitted in time for the trial on [the assignment due date].

4

Finding the Salami Curve

The year is 20xy. Hockey fans throughout Manitoba are following the fortunes of the Winnipeg Gliders, the city's entry in the new, co-ed Intercontinental Hockey League. But things are not going well for the Gliders; the team is mired in last place, game attendance and revenues are down, and the owners of the Gliders are threatening to move the team to Buenos Aires, which desperately wants an ICHL franchise.

One morning you are sitting in the office of Math Iz Us (the small consulting firm that you and your two partners have recently opened after having difficulty finding satisfactory summer jobs) when the phone rings. On the line is Jacques Schtrop, the Gliders' coach.

"I think that we may have the solution to our team's problems!" he tells you. "We've just signed the young European superstar Tina Salami, and she'll be joining the team next week. According to our scouts she has a slapshot that's been clocked at 135 miles per hour. Even if she's far from the goal when she shoots, the puck travels so fast that the goalie doesn't have time to react before it's in the net. And if the poor guy happens to be in the way of the puck, it knocks him right into the net and follows him in. We've never seen anything like it!"

"That's wonderful news!" you reply. "But how can I help you?"

"Tina has one little problem," continues Jacques. "She lacks accuracy. When she shoots you never know what direction the puck will go, so most of her shots aren't on goal. We've been working with her, and she's better than she used to be, but it would really help if she was subtending the largest possible angle to the goal-mouth when she shoots. That way the problem caused by her inaccuracy would be minimized. The trouble is, we don't know where she should be for that to happen."

"That's kind of vague," you say. "When you say 'largest,' what are you comparing the angle to?"

"Oh yes, I forgot to tell you about Tina's other problem," says Jacques. "She doesn't maneuver well on skates. She can skate very fast, but she can't change direction easily. So what happens is that she gets the puck at her own end of the rink and then she skates straight down the ice, along a path perpendicular to the goal-mouth, until she shoots. When she's far from the opposition goal she's subtending a small angle, and if she gets too close to the extension of the goal-mouth she's shooting from a very sharp angle, as we say in the trade. Somewhere along that path the angle that she subtends is a maximum, and we don't know where it is. **We need that information!**"

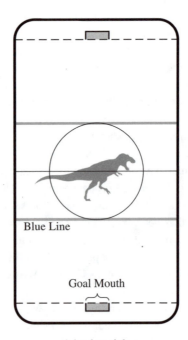

A hockey rink.

"Of course, if she were skating directly towards the goal-mouth, then the closer she is the bigger the angle she will subtend, so ideally she should skate right to the goal-mouth before shooting," you observe.

"Sure, I know that," says Jacques, "but if the path that she skates along doesn't intersect the goal-mouth then the answer's not so obvious. It seems like a mathematical problem to me, and that's why I've called you."

"It sounds like an optimization problem," you say, "but how far from the extension of the goal-mouth she should be for the angle subtending the goal-mouth to be a maximum surely must depend on the perpendicular distance of the skating path from the nearer goal-post. Look, Mr. Schtrop, let me talk it over with my partners and I'll get back to you."

"OK, but before you go, let me tell you what we would really like. We would like to get our icemaker to trace a faint green curve in the ice that would show Tina where to shoot. In other words, no matter what path she takes down the ice perpendicular to the goal-mouth, she should know that when she crosses the green line, she's at the best angle for shooting for the particular path that she is on. What we need you to do is to tell our icemaker how to make that curve."

"A 'Salami curve', so to speak," you say.

"Yeah, I guess you could call it that," says Schtrop.

When you discuss the problem with your partners, you realize that a knowledge of the Salami curve could be useful in sports other than hockey. Any sport played on a rectangular "field" requiring a player to shoot or toss a puck or ball into a goal would have the same feature. So, you decide to do a general analysis of the situation, in hopes of marketing the Salami curve to other teams besides the Gliders. Even your nephew who plays 10-year-old soccer for his community club might benefit!

Soon you realize that there are three relevant constants, namely the width of the playing surface, the length of the playing surface between the extensions of the goal-mouths, and the width of the goal-mouth. You assign letters to denote these. You also realize that you must consider each "path" perpendicular to the extension separately. Each path will contain an "optimal" point, and by stringing all these points together you'll get the Salami curve. Paths intersecting the goal-mouth are easy. When you concentrate on a fixed path not intersecting the goal-mouth, you are able to express the angle subtended as the difference of two inverse trig functions. Standard optimization techniques allow you to find the "Salami point" on that particular path, and combining them gives you the Salami curve. You write down the equation of the Salami curve using the three letters assigned above as parameters (in fact, it "comes in pieces"), and draw a very accurate, neat well-labeled diagram (on a large piece of graph paper) that is sufficiently clear for the icemaker to use as a blueprint for making the curve. Then you send the whole thing, together with an explanation of the mathematical derivation of the equation of the Salami curve, to Jacques Schtrop. He is delighted, and pays you a very large fee. You use the fee for a Hawaiian vacation, and pay no more attention to the Gliders for the rest of the season.

Epilogue   With the Salami curve available for all home games, Tina became more effective than ever. She scored 175 goals by the end of the regular season and the Gliders soared to second place overall. Fans attended in droves, profits were up, and the Gliders were Clinton Cup finalists, eventually losing to the Brisbane Emus on July 28 in the last game of a best of 15 series. Nonetheless, the Gliders did not stay in Winnipeg; Buenos Aires spent 150 million dollars constructing a new arena with 60,000 upholstered seats and an indoor roller coaster, and the owners moved the Gliders there in search of a larger profit. Winnipeg was forced to abandon its own tentative plans for a new arena, and spent the money thus saved on roads, sewers, libraries, and less severe tax increases.

Produce the package of materials that **Math Iz Us** mailed to Jacques Schtrop. Remember, in order to earn the full fee for the job, your work must be mathematically correct and complete, and explained clearly. Furthermore, the icemaker must be able to understand where to trace the Salami curve.

Those of you who are hockey buffs and who know the actual dimensions of a standard size NHL rink might determine whether the Salami curve crosses the blue line. If so, how far from the boards does this happen?

5

Saving Lunar Station Alpha

It was the loneliest summer job that Ranjana, Bruce, and Tara had ever had, and at the beginning they had wondered whether they should even take it. For four months they would be isolated in Lunar Station 50, so named because it sat astride the 50th parallel north of the equator of Xzyqgon, the small hollow spherical satellite of Neptune discovered in 2050. Lunar Station 50 had been abandoned in 2140, when the North American Federation had launched its eighth massive government downsizing and decided that the country could no longer afford to maintain it. However, more recently the Federation, concerned about possible moves by the Pan-Pacific Alliance to establish rival colonies, had decided to re-condition the old lunar stations and occupy them to prevent them from falling into unfriendly hands. Since high-paying summer jobs were no more plentiful in 2150 than they had been at the turn of the millenium, the three students, despite the hardship conditions, had decided to spend four months refurbishing the complex systems that regulated the heat, light, and air quality in the station. They had one further duty: they had to launch rockets from the old firing towers that still stood adjacent to the station. These towers were configured so that any rocket launched from them left the ground at an angle of 40° north of the vertical. They had been designed in this way so as to ensure that research rockets would travel on a path (essentially straight because of the weak gravity and lack of atmosphere of Xzyqgon) parallel to the axis of rotation of Xzyqgon and hence into the cloud of quasions that hovered above the north polar cap of the satellite.

All three of them wished that they had been chosen for the small number of highly coveted jobs at Lunar Station Alpha, the small city situated at the north pole of Xzyqgon. There, under its protective dome of superhardened transparent plexiglass, 50,000 happy citizens worked, played, and enjoyed the amenities that any contemporary city had to offer. Lunar Station 50, by contrast, offered only a dusty store of ancient interactive combat games in which the player could wrestle with a laser image of Arnold Schwarzenegger. These had little appeal for our heroes.

But the boredom that weighed heavily on the three students came to an abrupt end early in July when they received an emergency transmission from Lunar Station Alpha. "Two crises have arisen in quick succession," they were told. "First, our autotelescanner reported that a large meteor is approaching Xzyqgon; at 3:00 PM, three hours from now, it will be positioned at the zenith as observed from your location. We were able to determine its path, and it appears that it is moving in a straight line that will hit Lunar Station Alpha tangentially.

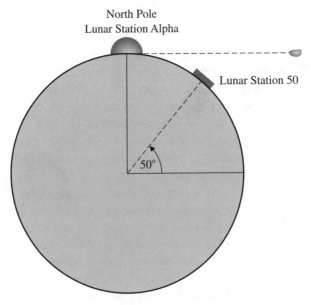

Cross-section of Xzyqgon.

"The second crisis is that before we could determine either its distance from us or its velocity, our entire computer system crashed. (Some things hadn't changed since the year 2000.) We can't count on our extra-Xzyqgon defense system to be up and running in time for us to launch an interceptor rocket, so we are going to need you to fire one of your quasion probe rockets loaded with a proximity fuse to intercept the meteor and blow it up into harmless small pieces before it strikes us. However, we will need to know whether your rocket will be able to pass sufficiently close to the meteor for the charge to detonate the way it is supposed to."

"Why can't we just hit the meteor with our rocket?" inquired Bruce. "Don't we have enough control over our rocket to be able to do that?"

"Unfortunately you don't," replied Alpha Mission Control. "Those old rocket launchers that you have cannot be aimed in any way—they have hardened positions pointing 40° due north of the vertical. Because the meteor will be at your zenith at 3:00 PM, in order to have your rocket's line of flight intercept the path of the meteor, you will have to fire at precisely 3:00 PM so that the meteor's path, your station, and the path of your rocket are all in the same plane. If you draw a circle representing the line of longitude that passes through both your station and our station, draw both the path of the meteor and the path of your rocket on your sketch, and indicate where the meteor will be at 3:00 PM, you will see what I mean."

"But with that configuration the path of our rocket will intersect the path of the meteor, so with the correct launch velocity can't we ensure that the two will collide?" asked Bruce.

"You could if you had complete control over the launch velocity of your quasion probe rocket. Unfortunately the maximum launch velocity that you have available to you is v km/hr, and we don't know whether that is fast enough for your rocket to reach the point of intersection of its path and the meteor's path before the meteor has already passed by. But that isn't essential; as long as your rocket comes within a certain fixed distance of the meteor, the proximity fuse will cause the explosives on the rocket to detonate, thereby shattering the meteor. We need you people to decide whether that will happen, because everyone here is engaged in trying to restore our computer system."

"But how can we do that?" asked Ranjana. "We don't have any of the relevant data! We don't know how far above us the meteor will be at 3:00 PM, we don't know what its velocity is, and I don't even remember what the radius of Xzyqgon is, and I think we're going to need to know that! And can we be sure that the meteor and our rocket will move at constant speeds in straight lines?"

"Remember that Xzyqgon is hollow, so its gravitational pull is so weak that flight paths are straight lines and speeds remain constant for all practical purposes," replied Mission Control. "We have the radius of Xzyqgon stored in our computer, and our Doppler radar has determined the speed of the meteor, so once we get the system working again we can beam it down to you. Then you can decide whether your rocket will ever get close enough to the meteor to detonate and destroy it. On that basis we can decide whether we need to evacuate the city. We want to do so only if absolutely necessary. As you know, survival outside our plexiglass dome is a touchy thing; anyone out there is susceptible to infection by the xzyqgoria virus, which is almost always fatal."

"So you want us to work up a mathematical model of the meteor-rocket system, and determine their closest approach in terms of our rocket's speed, the meteor's speed, the radius of Xzyqgon, and the distance of the meteor above us at 3:00 PM?" asked Tara.

"That's correct," replied Mission Control. "And then you can work out the numerical value of their closest distance apart once we have sent you numerical data for those quantities. Do you think you can do it?"

"Of course!" they responded in unison. "After all, we **did** study Intro Calculus!"

"Remember, though," added Mission Control, "although you have an upper limit for the velocity of your rocket, you don't **have** to launch it at that velocity, and it may be that the fastest possible speed is not the best one—you don't want your rocket to overshoot the meteor by a large distance and be long gone before the meteor makes it to the point of intersection of the paths. Or maybe not. Anyway, make sure that you write a detailed report for the archives explaining your methodology. We have a new Commander for the northern Xzqygon sector, Colonel Vance Frito the 7th, and he's a math freak. He'll want to see a complete analysis of the situation."

Prepare the detailed report that the three students will write for the archives. As you have no numerical data available, you will need to assign letters to the relevant quantities. What you need is an expression for the minimum distance between the meteor and the rocket in terms of the radius of Xzyqgon and the speed of the meteor. Neglect the component of the rocket's velocity imparted by the rotation of Xzqygon. Here are some specific questions to answer:

1. What must the maximum available velocity of the rocket be in order that the rocket can be made to hit the meteor?

2. If the maximum velocity of the rocket is only 98% of the velocity needed for collision, if the radius of Xzyqgon is 1000 km, and if the rocket must pass within 6 km of the meteor for the proximity fuse to explode the rocket and deflect the meteor, will our heroes be able to save LSα? Give your calculation.

6

An Income Policy for Mediocria

The summer of 20xy promised to be just as difficult as the previous summers for university students seeking employment. Heather, Sasha, and Li had approached hundreds of local businesses to attempt to convince them that they could benefit from the services of **Math Iz Us**. But replies were slow in coming, and the three students were beginning to be afraid that they would have to dissolve their consulting firm and accept jobs waiting tables at the campus restaurant instead.

As a last resort, Heather had suggested to the group that they place an ad in *The Campus Clarion*. So in the final week of March, next to the inquiries for lost wallets, parking lot passes, and tutoring requests, the following appeared:

> **Attractive Models Available** Do you have a complex financial or engineering problem that cries out for mathematical analysis? If so, look no further! **Math Iz Us** can provide you with the mathematical model that will let you solve your difficulties. Telephone xxx-yyzz.

Three days went by. Then late in the evening Heather received a phone call.

"Is this **Math Iz Us** ? Yes? Good. I saw your ad. My government needs the sort of assistance that you advertise, but as we are a very poor nation, we cannot afford the services of the usual high-priced consultants. However, I used to play pool at the Lizard Room, and I followed your contributions to the prosecution's case against Luigi. I was very impressed. I am returning home from my studies here tomorrow, and my father, who happens to be Vice President of Mediocria, has given me permission to hire you. If you choose to accept my offer, pack your bags for a long stay. The Government of Mediocria will be glad to be your host."

A discussion of the terms and conditions of employment followed. A quick consultation with her colleagues made it clear to Heather that Sasha and Li shared her views; although none of them could find Mediocria in an atlas, the pay was very attractive and the job (which apparently involved some sort of economic planning) sounded interesting. Furthermore, each of them relished the thought of travel to an exotic corner of the globe (even though globes don't have corners). So, two days later, they found themselves standing in the office of the Deputy Minister of Income Policy of the Government of Mediocria.

"I am very pleased that you accepted our offer," said the Deputy Minister. "My nephew enjoyed your country and your University, and even your pool halls. I am glad that he convinced my brother to offer you this job. We are short of mathematical talent in Mediocria, and I wish to have someone check our partial analysis of our income policy, and then extend it further.

"Here in Mediocria the government sets income policy. As everyone works for the government, we can decide what everyone earns. Of course, we are constrained by the total amount of money that we have available each year to pay our people. For sake of convenience, you may assume that we have a total of I Mediocrian dollars to pay out in salaries each year. Another thing that we do not control is the population of Mediocria. Our latest census indicates that the population is P. Of course, some are tiny children, and some are very old. Such people may get an allowance, or they may not receive any income at all. Now P is very large—many tens of millions. So, it is reasonable to use the methods of calculus in analyzing our income policy.

"Our problem is to fine-tune the amount that is spent by our population. Our empirical studies have convinced us that the fraction of one's income that one spends depends on the size of that income. The less one makes, the higher a fraction of it will be spent.

"Our one economics graduate who also knows some mathematics derived a model of spending behavior based on gauging the proportion of the population that make an income no greater than a fixed amount. More specifically, let A denote the average income (in our dollars) of citizens of Mediocria. (Clearly A is related to previously discussed quantities.) Let us suppose that we have decided that the highest permissible annual income in Mediocria shall be mA. As particularly productive citizens should be wealthy but not too wealthy, we might set m to be 2, or 10, or 50—that would be a policy decision. For each annual income level x, from 0 to mA, let $f(x)$ denote the fraction of the population that earns no more than x dollars per year. Our mathematical economist claims to have shown that

$$A = \int_0^{mA} xf'(x)\,dx\ldots, \tag{1}$$

but we aren't sure that he is correct. He told us that the vital observation was that the derivative $f'(x)$ was connected to the fraction of the population earning about x dollars per year. Can you let us know whether or not you agree with his analysis?"

The **Math Iz Us** trio retired to their workroom to mull over what they had heard. After an hour, Sasha broke the silence.

"I'm not sure if equation (1) is correct," he said, "but if it is, then given what we know about f, I can show, using integration by parts, that the difference D between the average income and the maximum income earned by anyone is given by

$$D = \int_0^{mA} f(x)\,dx. \tag{2}$$

"And I think that I can verify (1)," said Li, "if I start by trying to express the total income of all citizens of Mediocria by combining, for each income level, the contributions of all those whose annual income is at that level. Of course, that assumes that f is differentiable, and it's easy to imagine realistic situations where it won't be."

Elated by their success, the three consultants arrived at the office of the Deputy Minister bright and early the next day, and told him of their progress. "That is very good," remarked the Deputy Minister. "I can see that we were not mistaken in hiring you. Now comes the important part. Empirical study has shown that the fraction $r(x)$ of their income that a person receiving x dollars per year spends (rather than saves) is given by

$$\begin{aligned} r(x) &= 1 - (x/4A) \quad \text{if} \quad 0 \le x \le 3A \\ r(x) &= 1/4 \qquad\qquad \text{if } 3A \le x \end{aligned} \tag{3}$$

Thus a person with virtually no income spends almost all of it, while someone making the average income spends $3/4$ of it. We have decided that once we have chosen a suitable maximum income—in other words, once we have decided on a value for m—we want to have all income levels represented by the same number of people in the population. This means that we want $f'(x)$ to be a constant.…"

"Assuming that f is differentiable?" interrupted Li.

"Yes, assuming that f is differentiable," continued the Deputy Minister. "Then we want to see how the total amount of money spent by our citizens will vary according to the size of the maximum income we have set. Can you do that for me?"

"We can try," chorused the trio, and off they went.

"If $f'(x)$ is to be a constant," remarked Heather after some thought, "then the constraints that we have on f (which come from the definition of f), together with equation (1), allow us to determine a formula for $f(x)$ in terms of m and A, and furthermore we can show that this won't work unless the maximum permissible income is at least twice the average income."

"I've worked out a general formula, in terms of $r(x)$ and $f'(x)$, for the total amount spent annually by the citizens of Mediocria," added Sasha. "It's in the form of an integral. We can evaluate it explicitly for the particular functional form for $r(x)$ given to us by the Deputy Minister. From it we can see exactly how the spending rate will vary depending on the ratio of the maximum income to the average income. For the minimum possible value of this ratio, which Heather has shown is 2, we can get a numerical value for the percentage of all income that will be spent."

"And in fact," added Li after some scribbling, "it's not hard to show that no matter what that ratio is, at least 25% of all earned income will be spent each year. Of course, because of the form that $r(x)$ has, we need to consider two cases."

"Wonderful," said Heather. "Let's take it all to the Deputy Minister tomorrow morning."

When the trio walked into the Deputy Minister's office the next day, they were met by both the Deputy Minister and a tall, elegant woman. She sat silently as the trio gave a verbal description of their discoveries. Then the Deputy Minister spoke.

"Let me introduce you to the Right Honorable Vanessa Frito, our Minister of Income Policy. Until today I was not aware that Ms. Frito is a most sophisticated mathematician. She would like to make a personal detailed examination of your work."

"I'm not actually a professional mathematician, but I am a bit of a math freak," said Ms. Frito. "It runs in my family. What I need from you is a detailed written presentation of your results, with justifications for all the steps in your reasoning. When you have prepared a report that satisfies me, we will pay your fee and allow you to return to your homes."

It gets very hot and humid in Mediocria in the summer, and the country is plagued by large, aggressive biting flies. You want to prepare a report that will allow you to get out as soon as you possibly can. Your report should present all the results alluded to in the above narrative, and describe in detail how they were derived. You don't want Ms. Frito to require that you write a revised version.

[P.S. This project bears the same resemblance to real economics as "Saving Lunar Station Alpha" does to real engineering.]

7

The Case of the Cooling Cadaver

It was a dark and stormy night. The three members of **Math Iz Us** huddled in their sleeping bags and listened to the cold rain lashing against the outside of their tent. When Heather had first suggested it, the idea of using their very generous fee from the Government of Mediocria to finance a vacation spent cycling across Britain had appealed to them all. But the soggy British weather was not cooperating, and now they found themselves marooned in a tent on the grounds of a large old country estate. When they had arrived bedraggled at the door of the mansion, the elderly manservant who answered had taken pity on them—not enough pity to invite them in to warm dry beds, but enough to allow them to pitch their tent on the lawn a short distance from the building. "Lord Boddy is having some guests here this weekend," he quavered, "and there isn't any room, you know. Professor Prune and Colonel Catsup have the Regency bedrooms on the second floor, and Miss Carmine is occupying the vice-regal suite. There just isn't any room." The students accepted their fate stoically, and lay with their eyes open listening to the patter of the rain.

After an hour or so Sasha was roused from a light sleep by the sound of a heavy old door creaking on its hinges and then slamming shut. This was followed by the sound of jocular male voices and strange dull thuds. The rain had tapered off to a light drizzle, and Sasha could not contain his curiosity.

Glancing at his watch, which read 10 PM, he unzipped the door of the tent and looked out. Two men were engaged in a game of croquet. They seemed to be in high spirits, even though the rain soon began to come down harder. For a full hour they played, joked, and smoked enormous cigars, oblivious to the foul weather. Finally, just as the hands of Sasha's watch read 11 PM, a large chunk of masonry, loosened by the incessant rain, slid off the edge of the roof and landed with a wet thud on top of a recently played croquet ball. This seemed to startle the croquet players, and they vanished into the mansion.

Not more than a moment later the door opened and one of the croquet players re-emerged, accompanied by a young woman. They went into a gazebo, sat down, and began a low, animated conversation. By now Li was awake. "What's going on?" he asked Sasha. "For a wet night there's a lot of traffic outside," replied Sasha. "First there were two guys playing croquet, and now one of them is back with a woman." "Let's watch," said Li; and they did. But there was little to see; after an hour of earnest conversation the gazebo occupants abruptly got up and disappeared into the house at midnight.

Almost immediately thereafter a light switched on in a ground story room. Through the slightly open window came sounds of chatter punctuated by silence, together with clinking glasses. By now Heather was awake too. The rain had stopped, and all of them were restless, so they crept out of the tent and carefully made their way to a position just below the open window. Raising themselves carefully so as not to be observed, they looked in. A man and a woman were playing cribbage, drinking, and laughing. Sasha recognized the players as the woman in the gazebo, together with the croquet player who had not accompanied her to the gazebo. Crouching in cold wet bushes watching other people play cribbage is not an enjoyable way to pass time, and soon our heroes were back in their tent. But Heather could not sleep, and she lay listening to the sounds of the game until finally, at 1 AM (according to her watch), the noise ceased and the lights went out.

Almost immediately the trio was startled by the wail of sirens. Looking out of the tent, they saw that the mansion was ablaze with lights and police officers were swarming across the property. One of them approached the tent. "There's been an incident," he said. "You had better get dressed and come inside. You may need to assist us with our inquiries."

Inside they were led to a small wood-paneled room. Two sides held floor-to-ceiling bookshelves, while a third was lined with glass cases containing displays of insects. A large oak desk was pushed against the fourth wall. On the desk were a tensor light, a microscope, and some slides with mounted insects. The elderly manservant stood in the doorway wringing his hands. On the floor was the corpse of a large, white-haired man with a massive gash on the back of his head, surrounded by a pool of blood. A heavy brass candlestick lay nearby.

"As you can see, Lord Boddy has been murdered in the study with the candlestick," said an officer. "We don't know by whom. Our pathologist will be here soon, but in the meantime we must detain everyone."

"Perhaps we can help," said Li. "It will be important to know the time of the murder, and that can sometimes be determined by the temperature of the corpse. Is there a thermometer about?"

"I'll get one," said the manservant, and disappeared.

"Poor Sherlock," said the officer. "He's devastated. Old Sherlock Marples has been Lord Boddy's personal servant for the last 47 years. He's devoted his life to Boddy's welfare. And before you ask, no, he isn't a suspect. He's much too frail to deliver a blow like that. Besides, everyone knows he holds—held—Boddy in high esteem."

"I've checked the entire building," said another officer, poking his head in the door. "As Sherlock told us, all entrances are locked and there is no sign of forced entry. It must have been an inside job. Unless one of these...," he added, nodding towards the trio.

"No," said the first officer, "they aren't suspects. Sherlock says they were outside when he locked up, and only one door has been used since then. They had no access to it when the others were going in and out. There are only three suspects. We have them under guard in the drawing room."

"And those are..." began Sasha.

"Colonel Catsup, Professor Prune, and Miss Carmine. When you investigate a murder, you look for those with both motive and opportunity. Without going into the distasteful details, we happen to know that each one had a motive. And each is strong enough to have struck the fatal blow. So, we must decide who had the opportunity. Unfortunately there are no fingerprints on anything."

Sherlock returned with a rectal thermometer. "His Lordship always spent his evenings the same way," he said. "At 10 PM he would say goodnight, and then he would come here to his study to work on his collection of rare Peruvian beetles. Tonight I saw him enter the study myself, exactly at 10 PM. And then when I came by later and looked in because the light was still on, I found him. It's awful...."

"What time was that?" asked the officer. "About 1:00 AM," replied Sherlock.

"Perhaps I could have this," said Li, taking the thermometer. "Now let me see...." Selecting a convenient orifice, Li took the internal temperature of the corpse. "Let's see, it's 1:30 AM and the temperature is 32°C. Now we ought to be able to use Newton's law of cooling to determine the time of death. Because at that time, if Lord Boddy's physiology was normal.... "

"Oh, it was, it was!" interrupted Sherlock. "His physician always said that he was exactly normal in every way!"

"... was normal, then his body temperature was 37°C when he died, and began to cool according to Newton's law of cooling," finished Li. "So once we know what temperature the study is kept at, we can.... "

"Is there a wall thermometer in here?" asked Sasha.

"I'm afraid not," muttered Sherlock. "And in fact the temperature in this room seems to vary according to what direction the wind is blowing, what the outside temperature is... lots of things. This building is centuries old, you know. It takes a long time for the temperature to change, and in fact it seems exactly the same now as it was when I saw Lord Boddy into here at 10 PM."

"In that case everyone must leave immediately," exclaimed Heather. "We don't want the temperature of the study to be warmed by our collective body heat. We want it to stay exactly where it is. Otherwise the mathematics we're going to have to do will become impossibly complicated. Everyone out! But Li, you need to come back later and get some more data—preferably at well separated times.... "

"I'm sorry, ma'm," said the police officer, clearly impressed by the trio's knowledge, "but we can't detain the guests in the drawing room for too long. Already they are getting restless, and Colonel Catsup is threatening to call his lawyer. Our best hope is to determine the time of murder quickly, confront them with it, and see what happens."

"Can you wait for just two more hours?" pleaded Heather. "That way we can get two more readings separated by hour intervals—I hope that will be long enough—and do our calculations."

"I think we can manage two or three more hours," replied the policeman, "but that's the limit."

So the study lay empty and undisturbed for another two hours, except when Li took another pair of readings. At 2:30 AM the corpse's temperature was 30°C, while by 3:30 AM it had sunk to 28.25°C.

"Now comes the tough part," said Sasha. "We don't know when the murder occurred, although Sherlock has assured us that we do know Lord Boddy's temperature at that time. But we have three sets of readings, and three unknowns—the temperature of the study, the time of the murder, and the proportionality constant in Newton's Law. From what I know of matrix algebra, that should be enough."

"But we want to use some ingenious algebraic manipulation, not matrix algebra," said Li.

Each of the trio scribbled. When they compared notes and saw they had reached the same conclusion, they let out a sigh of relief.

"You young people have done an excellent job!" came a voice from just outside the door. "Allow me to introduce myself. I'm Inspector Vance McFrito from Scotland Yard. Now that you have the essential piece of the puzzle in place, please accompany me to the drawing room."

When they entered the drawing room, our heroes were startled to see the three house guests in formal evening dress. McFrito wore the same under his trench coat. Sherlock served glasses of port, and McFrito leaned against the mantlepiece, warming himself in front of a blazing fire.

"You are probably wondering why I asked you all here," began McFrito. "These young people have pinpointed the time of the murder. All we need to know now is who has no alibi for that time. First, please introduce yourselves."

When the guests did so, our heroes immediately recognized that Catsup and Prune had played croquet, while Catsup had conducted the tete-a-tete with Carmine in the gazebo. Carmine and Prune had been the cribbage players. When Heather announced the time of the murder, one of the guests leapt up. "Yes, I did it, and I'm glad! I was going to do it in the kitchen with the knife but I missed my chance—that's the way the dice roll. I wasn't counting on a trio of snotty math students messing up my plans."

"Take this pathetic creature away," said McFrito to his officers. "And, if you would be so kind," he added, turning to the students, "I would appreciate. . . . "

"A complete written report, including a detailed justification of the mathematical analysis that we used to determine the time of murder. Because, of course, you're something of a math freak," chorused the members of **Math Iz Us**.

"Why, yes," responded McFrito. "How did you know?"

1. Whodunnit? When?
2. Give a detailed explanation of how you arrived at your conclusion. All the mathematical reasoning should be presented and justified. Your explanation should be in the form of a report to McFrito.

[This project bears the same relation to real forensic pathology as "An Income Policy for Mediocria" does to real economics.]

8

Designing Dipsticks

The "Great White North" holds a romantic appeal for many people, and Anne was no exception. So when she saw a posting at the Campus Employment Center advertising a summer job at a remote fishing lodge in northern Manitoba, she immediately filed an application and was delighted when she was hired. She dreamed of a summer spent catching 10-pound trout, reading Robert W. Service poems by moonlight, and listening to the haunting howl of distant wolves.

Reality proved to be somewhat different. She spent her days cleaning one-pound trout, reading outboard motor repair manuals by flashlight, and listening to the maddening whine of nearby mosquitoes. Even when immersed in one of the many mindless repetitive tasks that filled her days, she was unable to think about anything more interesting than how tired she was and how much her insect bites itched... until one day when Joe Moosemess, the lodge owner, called her aside.

"We've got a problem with the fuel storage tanks, Anne," he began. "You know that lightning strike that the generator shack took last night? I think it fried a lot of the electrical gear. Anyway, all the fuel gauges on the storage tanks have packed it in. We've no way of telling how much fuel we have left in any of them. And that's a real problem for us, because we need to know how much to have the pilots bring in when they fly up from the south."

"What does this have to do with me?" Anne thought to herself. To Joe she said, "Gee, that sounds bad. What do you think you'll do?"

"Well, back before we had electrical fuel gauges, we used dipsticks to keep track of fuel levels, like they still use to check the oil level in your car. The fuel tanks had a hole right at the top, and you'd insert the dipstick vertically through the hole down into the tank until it hit bottom, and then you'd pull it out. The length of stick that was wet was the depth of the remaining fuel, and from that you could calculate how much fuel was left. But those sticks were calibrated in a funny way—I could never figure it out. And now we don't have them any more—we threw them away years ago when the electric gauges were installed. We never thought that we would need them again. But Anne," continued Joe, looking at her hopefully, "aren't you supposed to be a math major at that college you go to? Do you think you could make us some new ones?"

Two visions of the next few weeks flashed through Anne's mind. In the first, she continued to spend her mornings turning the clients' catches into a pile of fish fillets and a pile of fish guts, while the black flies waited until both her hands were occupied and then attacked. In the second, she sat in the air-conditioned insect-proof lodge office calculating the mathematics of dipsticks. She would have time for coffee breaks, and for her volume of Robert W. Service poetry, and Joe would be so impressed and grateful when she solved the dipstick problem that he would give her a raise and extra time off and....

"Sure I can!" she replied. "Take me to your tanks!"

It turned out that there were 10 different tanks of various sizes scattered around the grounds of the lodge. They came in only two shapes, however, namely right circular cylinders and spheres. The right circular cylinders were mounted in frames so that the plane of their circular ends was vertical. At the very top of the spherical tanks, and halfway along the top of the cylindrical tanks, there was a screw cap which, when removed, revealed a small hole into which a dipstick could be inserted. A slotted tongue of metal, through which a dipstick could be passed, ensured that any dipstick would be inserted vertically. This meant that the bottom of the dipstick would touch the bottom of the tank when fully inserted.

"Do you know the dimensions of these tanks?" asked Anne. "And are any two of them the same? You seem to have a bunch of different sizes."

"Yeah, all that information is in the documentation that came with them when we bought them," replied Joe. "We have that stuff back in the office. And you're right, they're all different sizes. We got a deal from a guy who was going out of business, so we took whatever he had. But they're all either spherical or cylindrical."

Back at the lodge office, Anne studied the papers that came with the tanks. There were five spherical ones and five cylindrical ones. The base radius and length of each cylindrical tank were listed, as were the radii of each spherical tank. Anne decided to work out a general formula that would apply to each cylindrical tank, and a different general formula that would apply to each spherical tank. That way she could feed the relevant data about each individual tank—the spherical radius, or the length and the radius of the circular base, as the case may be—into the appropriate formula when she dealt with a given tank. The dipstick reading would allow one to know the depth of the remaining fuel in the tank, and from that the total volume of fuel left could be calculated. Since spheres were solids of revolution, one calculation was straightforward. The other involved finding the area of the portion of a circle on one side of a chord, and here again Anne used calculus to obtain the answer. In each case her formula gave the volume of liquid in the tanks as a function of tank dimensions and the dipstick reading.

That evening, filled with a feeling of triumph, Anne accosted Kevin. This was Kevin's first year working at the lodge; he had finished high school only a few weeks ago. This autumn he would begin University. When they had met at the beginning of the summer, they had soon found that they shared interests in both mathematics and poetry. Anne was bursting to tell him what she had been able to do.

"I only wish that you understood calculus so that I could show you all the gruesome details," concluded Anne after describing the problem and her proposed solution.

"Next year I'll take it at university," said Kevin. "But you know, I don't think that you need calculus to work out the answer for the cylindrical tanks. You used it to calculate the area of part of a circle, but I bet that ordinary geometry and trigonometry are enough for that."

"Why don't you do it that way," suggested Anne, "and then we can compare our two answers and see if they are the same?"

"OK," replied Kevin. "But I'm convinced that calculus is important. There's no way I can see to get the answer for the spherical tanks just using ordinary geometry."

Afterwards, when Kevin and Anne compared notes, they found that their two answers for cylindrical tanks, obtained by different means, were in fact the same. Reassured, Anne started thinking about how to build the dipsticks. First, however, she showed her work to Joe.

"That's great, Anne," he said after listening to her explanation. "I don't pretend that I understand all the mathematics. But I remember how our old dipsticks worked, and I wonder if you could make yours work the same way. Our old dipsticks had a bunch of depth lines marked on them, and next to each line they'd written what fraction of a full tank was left when the liquid was down to that level. So on this cylindrical tank, for instance, you see that the total cubic capacity is 2.356 cubic meters—look, it's printed on the side of the tank here."

"Yes, this tank has diameter of one meter and a length of three meters, so—um—that's about right," said Anne, doing some quick arithmetic on her calculator.

"So, with our old dipsticks," continued Joe, "you would stick them in, and the top of the liquid would be at some particular mark—there were 32 equally spaced marks from top to bottom—and the mark would be labelled, oh, .431, say, so that would be the fraction of a full tank that was left. So you'd multiply 2.356 by .431, and get...."

"You'd get 1.015, about," said Anne, punching in numbers.

"...and so there'd be about 1.015 cubic meters of liquid left in the tank," concluded Joe. "Can you do that for me? Can you calibrate the dipsticks with the fraction of liquid left?"

"I think so," said Anne. "Let me go see."

Working from the formulae she had previously derived, Anne found that it was easy to write new formulae which gave the fraction of the capacity of the tank occupied by the remaining liquid as a function of the fraction of the dipstick length (the tank diameter in each case) that was wetted. These formulae, she noticed, had the advantage of being independent of the dimensions of the tank, and dependent only on the shape of the tank. Then she set about constructing dipsticks. For a spherical tank of diameter D she selected a thin straight strip of wood whose length was enough longer than D to provide a handle to grip the stick with. Then she divided the portion of length D into 32 equal segments, marked by lines. Using her formulae, she wrote by each line the fraction of liquid that would remain in the tank if the stick were wetted to that line. Once that first stick was made, she repeated the process for dipsticks for the other spherical tanks, noting with satisfaction that no further calculations with her formula were necessary. Then she repeated the process for the cylindrical tanks.

Anne's dipsticks worked marvelously. Joe was delighted, and provided her with a week's free lodging at the resort as a guest. Anne had time to write a lengthy parody of a Robert W. Service poem, which she later got published in the campus literary magazine. But on her last day at the lodge, Joe approached her in a panic. "We need help! Billy wanted to calculate how much fuel was left in the one meter radius cylindrical tank but he used the dipstick for the one meter radius spherical tank instead by mistake! And now we can't find either dipstick! All we know is that Billy concluded that the cylindrical tank was 0.232 full. Would that be right ?"

"No, it wouldn't," replied Anne. "But I can work out the right answer by using my formulae. All I'll need to do is use Newton's method to solve a cubic equation, and that isn't hard. Just be glad Billy didn't use the stick for the cylindrical tank on the spherical tank. That would have been a much harder problem!"

1. Joe has asked Anne to leave him a clear, complete description of how she derived her formulae, together with Kevin's alternate derivation for cylindrical tanks. Provide this writeup.

2. Make two model dipsticks as described above, one for a spherical tank of diameter 96 cm and one for a cylindrical tank of diameter 96 cm. Calibrate them as described. (You might want to start with a couple of meter sticks.)

3. What fraction full was the cylindrical tank on which Billy used the spherical tank dipstick? Provide the details of the calculations that Anne did to determine this. (Remember that the two dipsticks were lost at this point.)

9

The Case of the Gilded Goose-Egg

"Hello. You have reached the farm of Silas and Elsie Friday. We can't come to the phone right now, but if you leave us a message after the tone, we'll get back to you as soon as we can."

"Hello Mom, hello Dad, it's Della. I'm sorry that I haven't phoned you before now, but I've been really, really busy at work. Yes, I know you've been worried about me, and yes, I did get all those messages that you left on my answering machine. It's just that Per—that I ended up staying longer in the Bahamas than we—than I expected to, and when I got back to town I had to spend two weeks doing higher mathematics for a bunch of pirates. No, not the other lawyers, real actual pirates! I'm just so glad that Perry and I took that correspondence course in calculus after that case with the swiveling spotlight! Anyway, let me tell you about it.

"When we—when I got to the Bahamas just after Christmas, I just sat around on the beach for a few days and tried to forget about Lac du Portage and snow and everything. After that I got restless, and one day I found that I could take a trip on a sailboat out to a nearby reef and do some scuba diving. So early the next morning I was on my way for a day of sun and swimming. Little did I know!

"We spent all day diving and had loads of fun, and we were just starting back towards the resort around dusk, when this big speedboat came roaring out of the darkness towards us. It kind of cut us off, and we had to slow right down. Then a big guy climbed out on the deck of the speedboat and called over with a loud hailer. He wanted to know whether by any chance there were any lawyers who knew calculus aboard. Well, what could I say? Besides, I've always wanted to ride in one of those big speedboats. So Perry and I went over to the speedboat and asked if we could help.

"Well, Mom, can you believe it? Those guys on the speedboat were pirates! At least, that's what they told us. And they wanted us to help them divide their booty!

"Their captain told us that they had plundered this old Spanish shipwreck, and had recovered a really neat object. It looked like a huge medicine ball. It was perfectly spherical, and it was covered with gold! But the captain told us that they had been testing it, and it was really just lead plated with a thin veneer of gold that was the same thickness everywhere. Still, the sphere was so large that even though the thickness of the gold was very small compared to the radius of the sphere, there was enough gold there to be really valuable.

"Then the pirate captain told us why he needed us. Apparently it was the pirates' custom to divide all their booty equally, so they needed to know if there was an easy way to divide the sphere into pieces so that every member of the crew would get the same amount of gold. He wanted me to figure out a way to do this, and then draw up a legal document affirming that my proposed method of division would be fair. That's why he wanted a lawyer who knew calculus! For reasons of security he wouldn't tell me how many crew members there were. I thought that would make the problem really hard—but it didn't.

"Well, it didn't take me long. Since the thickness of the layer of gold was very, very small compared to the radius of the sphere, it was clear that if the sphere were divided up into a bunch of slices, the amount of gold on each slice would be essentially the outside surface area of the slice multiplied by the thickness of the gold veneer. Guess what, Mom! The solution is **really** neat! It turns out that if you have n crew members, and you slice the sphere into n slices of equal thickness, then every slice carries the same amount of gold, whether it's from the middle, or the end, or somewhere between! So all the pirates had to do was use their super-sharp diamond-toothed saws to slice the sphere into n pieces of equal thickness and their problem was solved. I got to keep the bit of gold dust produced by the saws as my commission.

"The captain asked me to write up my solution so his crew could read it. He told me that they had all done their graduate work at the Institute of Quantitative Piracy, so they all would be able to follow my solution and be convinced that things were fair.

"Once my solution was written up, I asked the captain if he would take us back to the resort. Not just yet, he said—and then we found out what was really on his mind. He waited until his crew was busy, and then he motioned me down to his cabin. But it turned out that he was still concerned about gold. He reached under his bunk and pulled out a giant gilded goose-egg—at least that's what it looked like at first sight.

"Apparently they had also recovered this at the same time that they had found the gilded sphere. It wasn't really shaped like a goose-egg—in fact, when we wrapped a rope tight around it and did some measurements, it became clear that it was an ellipsoid of revolution, the kind of solid that you get by rotating an ellipse around its long axis of symmetry. Just like the sphere, it had a lead center coated with a thin veneer of gold. The veneer was the same thickness all over, and that thickness was very small compared to the sizes of the axes of the ellipse.

"The captain then told me that he had known about the solution to the problem with the sphere all along, and that what we had just gone through with the crew was a charade to convince them that any time you had a problem like this, the fair way to divide the spoils was to saw the solid into slices of equal thickness. He suspected that this wasn't true for his goose-egg, but he wasn't really sure. He wanted me to find out whether or not everyone would get the same amount of gold if the goose-egg were sliced into n slices of equal thickness, and moreover, if they didn't, he wanted to know which slice would have the most amount of gold on it so that he could keep it for himself! He wouldn't even give me the dimensions of the axes of the ellipse—he said he wanted a general formula in case he ever got his hands on some other objects like that.

"Well, I sat down to do the calculations, and they became really messy. Finally I told the captain that I'd need more time, and that if I didn't get back to my hotel soon people would miss me and come looking for me, and that I promised to solve the problem for him as soon as I got back to Lac du Portage and mail him the answer. Well, he said he would let me go, but that he knew where I lived, and that I better have the problem solved in two weeks or he would ... Mom, I can't tell you what he said he would do.

"Anyway, once I got back to Lac du Portage I worked night and day. My biggest problem was that I thought I had to evaluate a very complicated definite integral (Joey can explain integrals to you). I could have done it if I'd had to, but it would have been rough. Then suddenly a few days ago I realized that if all I needed to do was to find the position of the slice with the maximum amount of gold on it (if there were such a slice), then I could use a really fundamental theorem of calculus and avoid having to evaluate the definite integral! After that things went quickly, although I had to be careful about being uncritical about critical numbers.

"So the solution went into the mail yesterday (along with my bill!), and now I can relax. Oh, say 'hi' to Joey for me. By the way, is he still looking for a summer job? The captain told me that he was hoping to hire a mathematically inclined young pirate, and I think that Joey would fit in really well with that bunch.

"Bye for now, Mom and Dad. I'll call you again soon."

1. Produce the analysis of the sphere that Della did while she was aboard the pirate ship.
2. Produce the analysis of the goose-egg that Della did when she was back in Lac du Portage.

Make sure your explanations are clear, accurate, literate, and complete, or else you may be keel-hauled.

10

Sunken Treasure

March 27, 20xy

Dear Mom and Dad,

I knew that you two would be worried about Joey and me, so I decided that I better try to have a letter smuggled out to you. The pirates have been **very** insistent that Joey and I won't be allowed to communicate with **anyone** until we finish the project. But one of the crew seems really nice, and I think he kind of likes me, so maybe I can persuade him to fax this to you. I sure hope so—I hate to think about you guys sitting at home fretting and not knowing what's going on. And could you let the other lawyers at my firm know that I won't be back to work until next week? Thanks.

Right now I bet you're saying to yourselves, "Why did Della have to go back to the Caribbean for a vacation again this winter, after what happened to her last year? Wasn't her adventure with the gilded goose-eggs enough for one lifetime? And why did she have to drag Joey along with her? Doesn't she realize that Joey should be back in class? Spring break is over!"

Well, folks, if I didn't have my baby brother with me, I'd be in even more trouble than I am now. After all, Joey knows more calculus than I do, and calculus is what the pirates need to solve their problem. Actually, it's quite an interesting problem. I only wish that we weren't going to be walking the plank if it's not solved in 48 hours... just kidding, Mom.

What happened was, Joey and I were walking along the harbor front in Maine d'Espagnol, which is this really picturesque town on the south side of the island, looking in the shops, when this big gruff guy stops in the street, stares at me, and says, "It's Della! Della Friday! The lawyer with the calculus!" At first I didn't recognize him without his parrot, but then I saw he was the pirate captain I had helped when I was down here before. He acted really glad to see me, and he told Joey and me that his ship was in harbor for a few days, and why didn't we drop by that evening for a few yo-ho-hos and a bottle of rum? Well, Joey was suspicious and didn't want to go, but you know us lawyers—we have this naive faith in human nature. So, off we went to his ship.

Well, they must have put something in the drinks, because when we woke up we were out to sea. The pirate captain didn't waste any time. He apologized for kidnapping us, and then he got right down to business. Apparently the pirate crew has salvaged this old hulk of a freighter barge. It has a really simple shape—its cross section is part of a parabola, with the cross section of the flat deck perpendicular to the axis of symmetry of the parabola. I'll draw you a rough sketch so you can see. The top of the barge is flat, with a hatch in the middle. The pirates have come up with this plan to use it to hide away all their stolen booty from the authorities. They intend to fill it with their treasure, pump its bilge tanks full of water, and sink it to the ocean floor. They are going to attach a small floating buoy to it with a rope, so they can locate it and recover

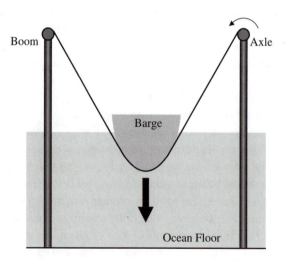

Face-on view of barge, boom and axle

it later once the heat is off them. Now the barge is very old and fragile, so they have to lower it very, very gently so that it's moving very slowly when it reaches the ocean floor. Otherwise it might break on impact and spill all the treasure out!

They have a neat plan for lowering it. They've taken over these two abandoned oil drilling platforms near here. Each platform has two vertical pylons jutting up, and the tops of all four pylons are the same height above sea level. The line segment joining the tops of the two pylons on platform alpha and the line segment joining the tops of the two pylons on platform beta form the opposite sides of a rectangle—sides that are about the same length as the length of the barge. The pylons are at the edges of their respective rigs, and there's nothing but open sea between them. The pirates have rigged up a horizontal boom joining the pylon tops on platform alpha. Between the pylon tops on platform beta they've set up sort of an axle that can rotate freely. It's attached to a powerful motor so that it can be turned slowly. They have this big roll of super-strong non-stretchy really flexible material wound on the axle, like a roll of film on a spindle. Once they've filled the barge with treasure, they'll maneuver it so that it sits exactly between the two platforms with its long axis parallel to the boom and the axle. Before they sink it, they'll unroll some of the material, pass it under the barge, and attach it to the fixed horizontal boom attatched to the top of the pylons on platform alpha, sort of like a sling for the barge. Then they'll fill the bilge tanks and slowly sink the barge. As it sinks, it will press down against the material from the roll. The pirates have rigged it so that the flat top of the barge will stay horizontal throughout the sinking process and the barge will move straight down, with no horizontal drifting. The motor will slowly turn the axle and pay out the material from the roll, allowing the barge to keep sinking further. Eventually the barge will reach bottom, the pirate will detach the roll from the boom and the axle, and let it sink too. Later they plan to return and raise the barge in the same way.

So, what do they want with Joey and me? They're not sure if their roll of material is long enough! They're scared that if there isn't enough material, then they won't be able to lower the barge gently all the way to the bottom, and it may break up if they just drop it the rest of the way. So they want us to figure out what length of material would need to be let out when the barge comes to rest on the ocean floor. They went ahead and took all the measurements that they thought we would need—the perpendicular distance between the axle and the boom, the height of the axle and boom above sea level, the depth of the ocean (luckily it's flat down there!), and the dimensions of the parabolic cross-section of the hull—both the length of the chord forming the top of the parabolic cross-section of the barge (that's the same as the width of the flat top of the barge) and the perpendicular distance from that chord to the vertex of the parabola (that's the same as the perpendicular distance from the flat top of the barge to the bottom of its hull). Joey and I think that's probably all the data that we'll need, and Joey has checked that the barge is sufficiently narrow to fit between the platforms. I wasn't sure about how to deal with the two places where the material in the roll first meets the hull, but Joey pointed out that the weight of the barge pressing down will pull things tight, so now I think we know how to handle it. Joey did say, though, that there are two cases to consider, depending on the relative sizes of the dimensions of the barge, pylon height and separation, and the ocean depth. But this darn captain! He thinks that he may want to do this same trick somewhere else where the ocean depth is different and the barge is a different size, so instead of just working out a numerical answer for this problem, he wants us to write a report assigning letters to the relevant dimensions and expressing the answer in terms of them. Then we'll compute the answer for the particular data we have here. I told Joey that it looks like we might have a messy integral on our hands and that it's a good thing that we have our calculators with us!

Anyway, I hear Joey calling—I think he wants me to check some of his calculations. I better go. I just **know** that you'll be **so** much more relaxed now that you've heard from us and know what's happening. Uncertainty is always the worst thing, isn't it? I'll call you the moment we get back. Keep calm. Joey says hello.

Love, Della

March 29, 20xy

Dear Mr. and Mrs. Friday,

I want you to know that we rescued your son and daughter unharmed from the clutches of the pirate crew, but not before they completed their work for them. Your children are real go-getters, and it wasn't a surprise to me when they told me that one of them is a lawyer and the other is an undergrad in the Business School at your university up there. Right now they're out trying to market their report to a local salvage company! They asked me to fax this letter to you and to let you know they are safe and will be home soon.

Sincerely,

Wyatt Cassidy
Chief of Police, Maine d'Espagnol

✤ ✤ ✤ ✤ ✤

1. Prepare the report that Della and Joey wrote for the pirate captain. Include a detailed description of your mathematical reasoning. Your goal is to find an expression for the length of the material that will be unrolled just when the barge first touches the bottom in terms of letters representing the distances that Joey and the pirates had measured.

2. Suppose the parabolic cross-section of the barge is 30 m wide at its widest point (at the top), and suppose that the cross-section is 27 m deep. Suppose the tops of the pylons are all 14 m above sea level, and that the horizontal distance from axle to boom is 80 m. What is the length of material that they will need if the ocean is 70 m deep at the place where the barge is being sunk?

11

The Case of the Alien Agent

It had been a long week for agent Muldie, and she was looking forward to a quiet Friday evening at home with a fascinating new book, "Conspiracy Theories throughout History." Unfortunately the New Age Network had postponed its special about how Atlantis was really a Martian colony peopled (aliened?) by pyramid builders who had fled to Egypt in an ark after a giant asteroid plunged into the Pacific and caused worldwide flooding, but her book would be a suitable substitute. As she read, though, she found herself growing drowsy, and soon the book slipped from her hands.

She was awakened by the ringing telephone.

"Muldie, it's Sculler. I've just received a very disturbing document in a plain brown wrapper. If what it says is true, it has major, major consequences for us all, but it's much too sensitive to discuss on the phone. Can you come over? After we go through it we can watch the latest episode of 'Occult Conundrums' together—I taped it yesterday. It's called 'Witchcraft in Wichita,' and my astrologer told me it was fascinating."

Witchcraft in Wichita—wow! And maybe agent Sculler still had some of that neat liqueur distilled from eye of toad and wing of bat! "I'll be right over!" Muldie replied. Only as she knocked on Sculler's door did it occur to her to wonder precisely what sort of disturbing documents Sculler was in the habit of receiving in plain brown wrappers.

"Look at this!" Sculler had spread the pages of a lengthy handwritten document across his coffee table. "Muldie, this is high-level, classified stuff. The letterhead says it's from the Institute of Paramilitary Paleogeology. I've never even heard of them, so they must be really supersecret. The person who sent this seems to be a disgruntled employee. He says that in his spare time he works for Alienwatch, the grassroots organization dedicated to exposing the truth about the aliens among us. He claims that the Institute of Paramilitary Paleogeology is suppressing the true facts about the amount of ash that fell to Earth after an ancient asteroid exploded in the atmosphere. He says that we owe it to the American people to expose the truth."

"An ancient asteroid?" asked Muldie. "You mean the one that's supposed to have crashed into the Earth 65 million years ago and wiped out the dinosaurs? The one that left a layer of iridium in the soil?"

"No, this guy's talking about an asteroid that he says exploded in the atmosphere 10 million years ago and wiped out the aliens that had colonized the planet and injected their DNA into the primitive primates of the time so that they would someday evolve into rational, scientific creatures like us. He says that the military and the high-level politicians know all about it, but they're suppressing it. He says that. . . . "

"Sculler, have you been into the batwing booze again? This sounds nuts to me."

"That's never stopped us before, Muldie. Here's his point. Apparently the military geologists have found a layer of soil like that iridium layer you were talking about. They've dated it to be about 10 million years old. It's composed of a hitherto unknown substance called freutcaquium—it's the fallout from the vaporized asteroid. Apparently it's toxic to non-carbon-based lifeforms. They claim that if there were enough of it, it would have poisoned the aliens and any Earth creature carrying alien DNA. It would have wiped them all out. But if there had been less of it, then some of the aliens might have survived and . . . their descendants might be living among us today! The truth is out there, Muldie, and we have to find it!"

"So what's in these documents?" asked Muldie, indicating the mass of paper spread over the table.

"Data and a calculation. The military geologists obtained core samples from widely scattered locations across the Northern hemisphere and found a very interesting variation in the thickness of the freutcaquium layer. Apparently it's thickest under that crater lake in northern Quebec, and then the thickness decreases linearly as you move away from 'ground zero,' so to speak, until

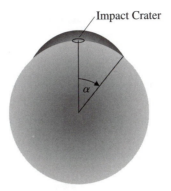

Cross-section of the freutcaquium layer (not to scale!).

beyond a certain distance from the crater there's none at all. It's an amazingly regular pattern, apparently."

"And so..." asked Muldie.

"And so that's enough data to allow the geologists to calculate the total volume of freutcaquium that settled to Earth. According to our informant, General Freutcaque sent..."

"Who?"

"General Frederic Freutcaque, the Director of the Institute of Paramilitary Paleogeology. The stuff is named after him. Anyway, he sent their calculation of the amount of freutcaquium to the Task Force on Alien Reproductive Biology so that...."

"To the what?"

"The TFARB is another of those government agencies that's so secret that no one I've spoken to has ever heard of it. That means that they are really, really important and sensitive. Apparently the TFARB decided that there was enough freutcaquium to render all the aliens sterile."

"And does your informant accept that answer? Assuming that any of this makes any sense whatsoever, that is."

"That's the point! The person who sent all this to me claims that the IPP calculated the volume of freutcaquium incorrectly! They forgot to take the curvature of the Earth into account, and they worked out the volume as though the shape of the freutcaquium layer was just a right circular cone. According to this guy, because they got the shape wrong, they might have overestimated the amount of freutcaquium that settled on the Earth. Do you realize what that means, Muldie?"

"Oh, yes! It means that you know that my calculus is stronger than yours, and you really invited me over here to help you to calculate the correct amount of freutcaquium."

Sculler reddened slightly. "Well... yes. That too. But Muldie... if their estimate of freutcaquium fallout is flawed... if it's too high... not all of the aliens, or primates with alien DNA, necessarily died! Some of them might have survived! We need to check the IPP's calculations. You could have alien genes yourself!"

"Well, I **am** feeling a bit alienated right now," retorted Muldie. "I didn't plan to spend the evening doing calculus. But sure, I'll help. Let's see the details."

"Unfortunately there isn't much numerical data in this material," said Sculler, turning to the documents on the table. "They've just set up a mathematical model. At ground zero the thickness of the freutcaquium layer is d meters, and from what they say it's clear that d is between 0.5 and 1. They took all these core samples for thousands of miles in all directions from ground zero, and they found that the thickness of the freutcaquium layer decreased linearly with distance from ground zero until it becomes zero at a distance of r kilometers from ground zero. From that they thought they could work out the total volume of freutcaquium fallout. But they goofed—they acted as though they were working on a flat surface."

"Wait a minute," said Muldie. "When you talk about distance from ground zero, you're talking about distance along the surface of the Earth, not along a straight line segment tunneling through the Earth—is that correct?"

"Yeah, that's right. If you want to be precise,...."

"Of course I do! We're doing mathematics here!"

"... distance is measured along the short arc of the great circle connecting two points. That's clear from what they write."

"Hmm... if they calculated their answer as though the Earth were flat, then the freutcaquium would be distributed in the shape of a right circular cone, and you could calculate its volume easily in terms of the parameters that you mentioned."

"And that's exactly what they did. But the Earth isn't flat—unless, of course, there's a conspiracy by the international astronomical community to keep the American people from knowing the truth about...."

"Sculler, stop it! Of course the Earth isn't flat! The question is, how much of a difference does it make to the answer that the Earth is a sphere?"

"That's why I asked you here, Muldie. How much of a difference **does** it make?"

✣ ✣ ✣ ✣ ✣

"I think I see what you're doing, but you better explain it to me," said Muldie, staring at a diagram on the blackboard in Sculler's study. It showed a picture of a semicircle, with the horizontal diameter along the bottom, surmounted by a thin crescent-shaped region, symmetric about a vertical line through the center of the semicircle, widest at the top of the semicircle and tapering to nothing partway down each side. The crescent-shaped region was shaded.

"If you look at this long enough you can see that the freutcaquium layer can be thought of as a volume of revolution swept out by rotating the shaded region," said Sculler. "Those parameters r and d that I mentioned earlier measure these distances on the shaded region." He scribbled some symbols on the board.

"Of course, there's one more parameter that's important that you haven't included—the radius of the Earth," observed Muldie. "Why don't we denote that by R?" She took the chalk and wrote on the appropriate place on the diagram.

"OK, now what I tried to do was to use the cylindrical shell method that they taught us in calculus," said Sculler. "What I got was a mess—the integral for the volume was so messy that I backed right off. I stared and stared, but I wasn't getting anywhere, so . . . . "

"So you called me, right?" replied Muldie. "I guess I should be flattered. Now I suppose you expect me to have a bright idea or something."

"That would be nice."

"First let me try to do it using cylindrical shells. It's not that I don't believe you when you say it's a mess, it's just that I like to see these things for myself."

Some minutes passed.

"First of all, it's pretty clear that we should be using the angle subtending distance along the Earth's surface as a variable," said Muldie. "We can relate that to distances along the Earth's surface by using R."

"I tried that too, but it didn't help," said Sculler. "The integral's still a mess." More minutes passed.

"You're right!" Muldie finally said. "I wonder if there's any way we can do this approximately without losing too much accuracy."

"Well, d is a meter at most, and the radius of the Earth is huge in comparison to that. Maybe that will help," suggested Sculler.

"Wait a minute!" exclaimed Muldie. "Maybe cylindrical shells aren't the thing to use after all!"

"You think we should be using discs?" queried Sculler. "That's the only other thing that my old calculus text suggested, and they're even worse than shells."

"No, something else," said Muldie. "Why not rotate little rectangles sticking straight out of the semicircle? They'll generate a shell whose surface area we can find, and then multiplying by the shell thickness will give us an element of volume."

"But won't the shell thickness vary?" asked Sculler.

"I guess it will," replied Muldie. "The question is, by how much?"

"Actually, that's where we can use the fact that d is very small compared to the radius of the Earth!" exclaimed Sculler. "To a high degree of accuracy that will allow us to assume that the shell thickness is constant."

"You're right!" replied Muldie. "And besides that, we can use the small size of d in comparison to R to get an accurate, simplified approximation for the surface area of that shell. We're in business!"

Ten minutes later they gazed proudly at the blackboard, which was now covered with a moderately lengthy computation culminating in a compact expression for the volume of freutcaquium. Their answer was expressed in terms of the angle α subtended by an arc of length r, the distance along the Earth's surface from ground zero to where the thickness of the freutcaquium layer reached zero.

"The question is, how much does our answer differ from the answer that the IPP people got?" asked Muldie. "Intuitively you'd expect that the bigger α is, the more effect the curvature of the Earth will have and the bigger the discrepancy there will be between our answer and the IPP's answer."

"I have an idea," said Sculler. "Let's write out the Maclaurin series for the trig function in our answer, and then do some simplifications. That will make it clearer to us what the difference is between the two results. Why don't you do that, Muldie, while I rummage through all this stuff"—he gestured at the messy manuscript on the table—"and see if I can find any hint about what α actually is."

Each of them worked silently. After some time Muldie said, "Using what I know about the size of the sum of an alternating series, I've gotten an upper bound for the percentage difference between their answer and ours that's in the form of a constant multiple of α. Sculler, could you find a precise value for α from their data?"

"It's pretty clear that α can't be any larger than $\pi/4$," replied Sculler.

"Well, in that case," said Muldie, "the IPP's answer is within... let's see, here... about 3% of the correct answer. Neglecting the curvature of the Earth didn't make enough difference to the IPP's answer to matter. So, even if all that garbage about aliens being here 10 million years ago is correct, if the TFARB is right about the amount of freutcaquium that it takes to kill an alien, there wouldn't be any left. There are no alien genes in me after all, Sculler."

"But that can't be correct!" exclaimed Sculler. "I **know** that some creatures with alien genes survived!"

"Nonsense! How could you possibly know that?"

"Because," he intoned, "I'm one of them. We aren't extinct . . . we roam the Earth!"

"Oh, my goodness!" gasped Muldie, backing away from Sculler, whose eyes had acquired a strange green glow. "Keep away . . . . "

As she backed up, her legs caught the coffee table, causing her to stumble backwards onto the floor . . . where she awoke on the floor of her own apartment.

It had all been a dream! She had fallen asleep and then rolled off her sofa.

"Of course it was a dream," she said to herself, "Why couldn't I tell that? After all, none of it made any sense. Mind you, that was a really interesting calculus problem. I do calculus asleep better than I do it when I'm awake!"

Her ruminations were interrupted by the telephone. She lifted the receiver.

"Muldie, it's Sculler. I've just received a very disturbing document in a plain brown wrapper. If what it says is true, it has major, major consequences for us all, but it's much too sensitive to discuss on the phone. Can you come over? After we go through it we can watch the latest episode of 'Occult Conundrums' together—I taped it yesterday. It's called 'Witchcraft in Wichita,' and my astrologer told me it was fascinating."

Write the report that Muldie and Sculler might have prepared for Alienwatch giving their computation of the amount of freutcaquium. Indicate why their approximations are valid, and by what percent their estimate of the amount of freutcaquium differs from that prepared by the IPP. (Express your answer in terms of α.)

Solutions

1

The Case of the Parabolic Pool Table—Solution

To: Judge Vance Frito
From: Math Iz Us
In the matter of: The People vs. Mr. Luigi McTavish

Your Honor,

In this brief we shall show that Mr. Luigi McTavish did keep in the establishment known as Luigi's Lizard Room, owned by Mr. MacTavish, two pool tables of similar design but different dimensions to aid and abet a fraudulant gambling scheme. We shall further show that the table used by Mr. McTavish was designed so that McTavish could win a bet with his co-gambler, while the table used by the co-gambler was designed to render it impossible for said co-gambler to win. This submission shall constitute evidence corroborating the charge of fraud against Mr. McTavish.

Consider a pool table constructed so that three of its sides comprise three of the four sides of a a rectangle, while the fourth side has been replaced by a portion of a parabola bulging outwards, a parabola whose axis of symmetry is the perpendicular bisector of the opposite side of the rectangle (see Exhibit A). This implies that the vertex of the parabola must be at the intersection of the parabola with said perpendicular bisector.

A cue ball is placed at the midpoint of the "missing" fourth side of the rectangle, denoted by an 'x' on Exhibit A. The cue ball is to be shot against the parabolic portion of the cushion. The gambler wins if the cue ball returns over the x from which it was shot. We shall consider what the relative dimensions of the rectangular and parabolic portions of the pool table must be in order for this to be possible.

Let R denote half the width of the pool table (see Exhibit A; all distances will be measured in cm). Let L denote the distance from the x to the vertex of the parabola. Consider a set of coordinate axes whose origin O is at the x (where the cue ball is placed) and whose X-axis is the extension of the "missing" side of the rectangle comprising part of the pool table. Then the Y-axis is the extension of the axis of symmetry of the parabola, and we declare that the Y-coordinate of the vertex is to be positive. Then the corners of the table where the rectangular sides meet the parabolic cap have coordinates $(-R, 0)$ and $(R, 0)$ respectively, and the vertex of the parabolic cap has coordinates $(0, L)$. Now a parabola whose axis of symmetry is the Y-axis has an equation of the form

$$y = ax^2 + bx + c, \tag{1}$$

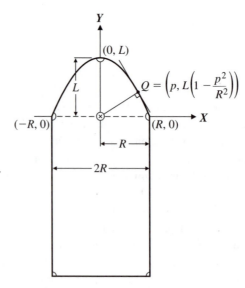

Exhibit A. The case of the Parabolic Pool Table.

where a, b, and c are constants. Substituting the coordinates of the points $(-R, 0)$, $(R, 0)$ and $(0, L)$ into equation (1), we obtain the equations

$$L = c,$$

$$0 = aR^2 + bR + c,$$

and

$$0 = aR^2 - bR + c.$$

Subtracting the third equation from the second and then dividing by the non-zero quantity $2R$ gives $b = 0$. Thus $0 = aR^2 + L$, whence $a = -L/R^2$. Thus using (1) the parabolic cap has equation

$$y = L - \left(\frac{L}{R^2}\right) x^2. \tag{2}$$

If the cue ball is to bounce back directly over the 'x' from which it is shot then, according to the standard "laws of bouncing," if the cue ball has no spin, its original path must be perpendicular to the cushion at the point where the path meets the cushion. More precisely, the original path must be perpendicular to the tangent line to the parabolic curve of the cushion at the point where the path meets the cushion. We now investigate under what circumstances this can happen.

Suppose that the cue ball hits the cushion at a point Q on the cushion whose X-coordinate is p (see Exhibit A). As there is a pocket at the vertex of the parabolic part of the cushion (i.e., at the point on the parabolic part whose X-coordinate is 0), we must have $p \neq 0$. By the symmetry of the situation we can assume that $p > 0$, but clearly $p \leq R$ as all points on the parabolic part of the cushion have X-coordinates between $-R$ and R (see Exhibit A). Using equation (2), we see that the Y-coordinate of Q is $L - (L/R^2)p^2$. We now derive two expressions for the slope of the path OQ followed by the cue ball.

Since $(0, 0)$ and $(p, L - (L/R^2)p^2)$ both lie on OQ, the "rise over run" formula for the slope of OQ gives:

$$\text{slope of } OQ = \frac{L - \left(\frac{L}{R^2}\right)p^2}{p}$$

$$= \frac{L(R^2 - p^2)}{pR^2}.$$

(Recall that $p \neq 0$ so this expression, which involves division by p, makes sense.) Second, by differentiating equation (2) we see that the slope $y'(p)$ of the tangent to the cushion at the point Q is $-2Lp/R^2$. The slope of the normal line OQ to the cushion at point Q is the negative reciprocal of this, i.e., $R^2/2Lp$. Equating these two expressions for the slope of OQ, we get

$$\frac{L(R^2 - p^2)}{pR^2} = \frac{R^2}{2Lp}.$$

Solving this for p^2 yields:

$$p^2 = R^2\left(1 - \frac{R^2}{2L^2}\right).$$

As $p > 0$ it follows that

$$p = R\sqrt{1 - R^2/2L^2}.\ldots \tag{3}$$

If $1 - R^2/2L^2 > 0$, then as $R^2/2L^2 > 0$ (as perfect squares are positive and $R \neq 0$), it would follow that $0 < 1 - R^2/2L^2 < 1$ and hence $0 < \sqrt{1 - R^2/2L^2} < 1$. Consequently by equation (3), it would follow that $0 < p < R$, and hence there would be a spot Q on the cushion from which the cue ball would bounce directly back over the 'x' as required; the X-coordinate of Q would be given by equation (3) above. It is clear from the symmetry of the situation that the point on the parabolic part of the cushion whose X-coordinate is $-R\sqrt{1 - R^2/2L^2}$ would also be such a spot.

If $1 - R^2/2L^2 = 0$ then by equation (3) $p = 0$ which, as observed above, is not a viable solution.

If $1 - R^2/2L^2 < 0$, then there are no real numbers p satisfying equation (3) and so there would be no spot on the cushion from which the cue ball could bounce back over the 'x' from which it was shot.

In summary, the cue ball can be bounced off the cushion (without spin, so that the "bouncing laws" apply) so that it will pass back over the 'x' if and only if $1 - R^2/2L^2 > 0$. Rewriting this inequality, we see that this happens only if $R/L < \sqrt{2}$.

When we measured the tables used by the defendant and the co-gambler, we found that $R/L = 1.31$ for the table used by the defendant, while $R/L = 1.58$ for the table used by the co-gambler. We consequently conclude that the defendant Mr. McTavish was operating a gambling game in which the co-gambler had no chance of winning, but in which Mr. McTavish could win with the exercise of a normal degree of pool-playing skill.

2

Calculus for Climatologists—Solution

When Anne arrived at Thirsty's the following Friday afternoon, she saw that Len and Jason had gotten there before her and had chosen a quiet table near the windows. Len was slouched back in his chair, absorbed in the demanding task of cleaning his fingernails with the blunt end of a toothpick. Jason was systematically tearing a paper napkin into thin strips. As there was no food or drink on the table, they apparently had not yet been served.

Jason looked up. "What happened to you?" he asked, giving Anne a rather fierce look. "You're late."

"I'm sorry," replied Anne. "I wanted to type out the solution and print it before I came here, and the printer jammed, and it took about 20 minutes to get it cleared. Anyway," she continued, opening her backpack and pulling out a sheaf of papers, "here it is." She gave several sheets to each of the others.

Len stirred slightly. "So, you've actually got a solution? Good for you! What's the answer?"

"You were right, Len. Any great circle has to contain two antipodal points at the same temperature. I've got the proof right here."

"You actually proved that?" exclaimed Jason. "This I've got to see. You know, you had the easy job—if I was going to prove that it it couldn't happen, I'd have had to have temperature readings from every single point on some great circle at the same instant of time, and there just isn't the available meteorological data to do that, and I bet there never will be!"

"Well ... Anne's going to show that it always happens on every great circle at every instant, and she has no more access to weather data than you do, Jason," replied Len. "Showing that something has to happen isn't the same as giving an example of it happening. Mathematics is full of what are called existence proofs, where you know that something has to be there—like your antipodal points of equal temperature—but you don't know where or what."

"Can I show you how I did this?" asked Anne. "I want to know if you agree with my reasoning."

"Go ahead," said Jason. "I'm listening."

"OK, I'm going to show you how this works if the great circle is the Equator, because it's easier to understand if you're talking about a particular circle, but I think you'll see that the same argument works for any great circle."

"That sounds like cheating to me," replied Jason.

"Just listen first, OK?" retorted Anne. "I can do it for any great circle, but if you are as smart as you always say you are, Jason, I won't need to. So just listen. Anyways, what I did is use the Intermediate Value Theorem—let's just call it the IVT—for continuous functions. Len told us last week that the temperature function was continuous, so that's what gave me the idea."

"So you applied the IVT to the temperature?" asked Len.

"No, it wasn't quite that simple," replied Anne. "I built a continuous function out of the temperature function and applied the IVT to that. Suppose we define the temperature function T on the equator by letting $T(x)$ denote the temperature at a particular instant at the point on the Equator that is an angle of x radians to the west of the point on the Equator that's at longitude $0°$."

"But if x is bigger than 2π, then you go back over the same points again," said Jason.

"That's right, and we get the same point when x is 2π as we did when x is 0," replied Anne. "So think of the domain of T as being $[0, 2\pi]$ and just notice that $T(0) = T(2\pi)$. Now, as Len remarked, T is a continuous function. Let's define another function a with domain $[0, \pi]$ by saying that $a(x) = x + \pi$. It's pretty clear that a is continuous."

"Even I could prove that," said Jason.

"Now let f be defined by the equation

$$f(x) = T(x + \pi) - T(x),$$

and give f the domain $[0, \pi]$. Notice that if $x \in [0, \pi]$, then $x + \pi \in [\pi, 2\pi]$, and so the definition makes sense. In functional notation we can write that $f = T \circ a - T$ and, because we know that compositions and differences of continuous functions are continuous, we know that f must be continuous. What we're going to do is apply the IVT to f on $[0, \pi]$."

"Cool," commented Len.

"Now, observe that

$$f(0) = T(\pi) - T(0),$$

and

$$f(\pi) = T(2\pi) - T(\pi)$$
$$= T(0) - T(\pi)$$
$$= -f(0),$$

so $f(0)$ and $f(\pi)$ have opposite signs—one is positive and one is negative," continued Anne. "The IVT says that if a function g is continuous on a closed interval $[a, b]$ and if d is any number strictly in between $g(a)$ and $g(b)$, there is a number c in between a and b with $g(c) = d$. So here, since 0 is strictly between $f(0)$ and $f(\pi)$, there is a number r between 0 and π such that $f(r) = 0$. This says that $T(r + \pi) = T(r)$, so the points on the Equator that are r radians and $r + \pi$ radians west of longitude zero are our antipodal points of the same temperature."

"You forgot one thing," said Jason. "Suppose that $f(0) = 0$? That's a possibility that you didn't consider."

"In that case we have $T(\pi) = T(0)$, and so the points on the Equator at $0°$ longitude and $180°$ longitude (i.e., π radians longitude) are the antipodal points of equal temperature. See, it always works!" finished Anne triumphantly.

"Yeah, I guess it does," said Jason. "And not only do you not have to convince me that your argument adapts easily to other great circles, but it can be adapted to other circles on the surface of the Earth, like the Tropic of Cancer. It's not a great circle, but if you define T and f exactly the same way that you did for the Equator, your argument will work the same way."

"It goes further than that," said Len. "You can find two antipodal points A and B on the boundary of Colorado that are at the same temperature right now. Of course, here we need to define 'antipodal' to mean that whether you go from A to B along the boundary in a clockwise or counterclockwise direction, it's the same distance. In fact, the argument works for any simply closed curve on the surface of the Earth."

"What's a simply closed curve?" asked Anne.

"Ah, but that's another story," replied Len. "Waiter, bring this woman a drink. Anne, what will you have?"

"A small ginger ale, thanks," said Anne.

3

The Case of the Swiveling Spotlight—Solution

Your Honor,

The defense proposes to show that if the defendant were traveling at the speed that the police witnesses claimed to have measured, then he could not have been under observation for as long as the 20-second minimum that the law requires. We will establish this by deriving an equation whose root T is the time (in seconds) that would have elapsed from when the defendant's car was first observed until the time when the spotlight on the police cruiser would have been pointing directly sideways. According to the testimony from Sergeants Preston and Renfrew, T is larger than the time that the defendant's car was actually under observation.

We will derive a series of approximations to T. We will show that each of these approximations is larger than T, and that one of them is less than 20. From this we will conclude that T is less than 20, and hence that the defendant's car was under observation for less than the 20 seconds required by law.

Consider a system of coordinates imposed on a map of Park du Portage and its surroundings (see Exhibit A). The X-axis is tangent to the parabolic boulevard at its vertex, and the Y-axis is the north-south line passing through

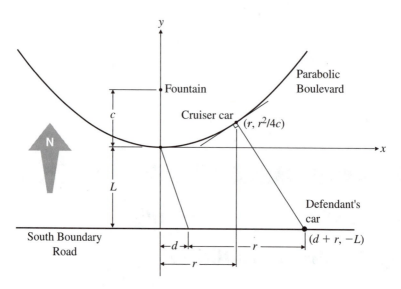

Exhibit A. The Case of the Swiveling Spotlight.

the vertex of the boulevard. Hence the origin is at the vertex. The south boundary road, along which the defendant's car traveled, now forms a horizontal line below the X-axis. The fountain in Park du Portage is at the focus of the boulevard, directly north of the vertex as shown on the map, on the positive Y-axis.

The following distances have been measured (in feet):

- Let c be the distance from the origin to the fountain. Hence the coordinates of the fountain are $(0, c)$.

- Let L be the perpendicular distance from the vertex (i.e., the origin) to the south boundary road. Thus the equation of that road is: $y = -L$.

- Let d be the east-west separation between the police cruiser car and the defendant's car. Officers Preston and Renfrew have testified that they have measured this distance, and that it remained constant throughout the tracking of the defendant's car. The coordinates of the defendant's car when it was first observed were $(d, -L)$.

- Let v be the purported speed at which the defendant was traveling (in feet/second), and let r denote the distance he would have traveled from when he was first observed until the spotlight would have been perpendicular to the direction of travel of the cruiser car. If T was the time elapsed, it is well-known that

$$r = vT. \qquad (1)$$

So, when the spotlight was perpendicular to the cruiser, the defendant was r feet due east of his original position; hence his coordinates at that moment were $(d + r, -L)$.

Your Honor, it is well-known (see any treatise on conic sections) that the equation of a parabola opening upwards, with vertex at the origin and distance c from vertex to focus, is

$$y = \frac{x^2}{4c}. \qquad (2)$$

Hence the coordinates of the cruiser after it had travelled r feet east were $(r, r^2/4c)$.

We will now derive two expressions for the slope of the line traced by the spotlight beam pictured in Exhibit A. Equating them will yield a cubic equation for T.

The spotlight beam pointed directly sideways from the direction of the cruiser car. Since the cruiser car was moving in the direction of the tangent to the parabolic boulevard, the beam formed a normal line to the parabola at the point $(r, r^2/4c)$ occupied by the cruiser. The slope of the tangent line to the boulevard at the point $(r, r^2/4c)$ is

$$\left.\frac{dy}{dx}\right|_{x=r} = \left.\frac{2x}{4c}\right|_{x=r} = \frac{r}{2c}.$$

As the product of slopes of perpendicular lines is -1, the slope of the spotlight beam is $-2c/r$.

Since we know the coordinates of two points (i.e., the cruiser car and the defendant's car) on the line traced by the spotlight beam, we can use the "rise over run" formula for the slope of a straight line to express that slope as

$$\frac{\frac{r^2}{4c} - (-L)}{r - (d + r)} = \frac{r^2 + 4cL}{-4cd}.$$

Equating our two expressions for the slope of the spotlight beam, we obtain

$$\frac{-2c}{r} = \frac{r^2 + 4cL}{-4cd},$$

which simplifies to

$$\frac{r^3}{4c} + Lr - 2cd = 0.$$

Replacing r using (1) we obtain

$$T^3 + \left[\frac{4cL}{v^2}\right] T - \frac{8c^2d}{v^3} = 0. \tag{3}$$

Let $A = 4cL/v^2$ and $B = 8c^2d/v^3$. Then $A > 0$ and $B > 0$ and (3) becomes

$$T^3 + AT - B = 0. \tag{4}$$

To analyze this, let $f(t) = t^3 + At - B$. Then $f'(t) = 3t^2 + A \geq A > 0$, and so $f'(t) > 0$ for all $t \in \mathbf{R}$. Thus f is increasing on $\mathbf{R}$, and so (4) can have at most one real root. However, straightforward calculation shows that $f(0) = -B < 0$ and $f(B/A) = (B/A)^3 > 0$, so as f is continuous (being a polynomial), the Intermediate Value Theorem and (4) tell us that $0 < T < B/A$.

Observe that $f'(t) = 3t^2 + A$ so the minimum slope of the graph of f is A, which is positive. Also, $f''(t) = 6$ so the graph of f is concave up on $[0, \infty)$. Hence the graph of f has the shape indicated in Exhibit B.

We now use Newton's method to approximate the (unique) zero of f. In general, one makes a guess (call it t_1) as to what the root of $f(t) = 0$ is. One then draws the tangent to the graph at $(t_1, f(t_1))$ and lets the second approximation (call it t_2) to the root of the equation be the first coordinate of the point where the tangent line meets the horizontal axis. As you can observe in any calculus text, your Honor, one finds that

$$t_2 = t_1 - \frac{f(t_1)}{f'(t_1)}. \tag{5}$$

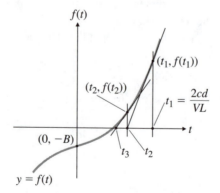

Exhibit B. The Case of the Swiveling Spotlight.

One then iterates this procedure. In general the $(n + 1)$st approximation t_{n+1} to the root of the equation is obtained from the nth approximation t_n by the equation

$$t_{n+1} = t_n - \frac{f(t_n)}{f'(t_n)}. \tag{6}$$

We saw that $T < B/A$, so we let our first approximation t_1 be B/A, which equals $2cd/vL$. Since the graph of f has positive slope and is concave up on $[0, \infty)$, we see from Exhibit B that $T < t_{n+1} < t_n < B/A$ for each positive integer n. We can use (6) iteratively to compute each t_n in terms of the quantities c, d, v, and L (where v, of course, is the speed at which the police claim my defendant was driving).

When I obtained the numerical values for c, d, v, and L and used them to compute the value of t_n for successive values of n, I found that although t_1 and t_2 both were larger than 20, t_3 was less than 20; consequently T was less than 20. The law therefore requires that my client be acquitted.

Respectfully submitted,
Ms. Della Friday

4

Finding the Salami Curve—Solution

To: Mr. Jacques Schtrop
    Coach, Winnipeg Gliders

Please find below our report giving the equation of the Salami curve relative to a suitable coordinate system, together with a scale diagram. We have included our mathematical derivation of the equation of the curve. In order to accommodate the different sizes of rinks found in the ICHL, we have denoted relevant distances with letters; you can substitute the appropriate numerical values for your rink. We trust that this is satisfactory.

Sincerely,
Math Iz Us

Let L denote the perpendicular distance between the goal-mouths. (All distances are measured in feet.) Let w denote the width of the portion of the rink between the goal-mouths. Let d denote the width of the goal-mouth. As the midpoint of the goal-mouth is equidistant from the two sides of the rink, we obtain a diagram as seen in Figure A.

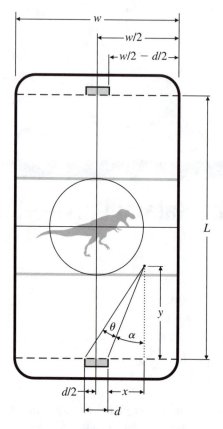

Figure A.

Suppose that Tina skates along a path to the right of the right-hand goal-post (indicated on Figure A by a broken line) that is perpendicular to the extension of the goal-mouth, and suppose the perpendicular distance from that path to the nearest goal-post is x (where $0 \leq x \leq (w - d)/2$). When Tina is a perpendicular distance y from the goal-mouth she is approaching, let her subtend an angle θ with the goal-mouth. Let the angle between the path she is following and the line from her to the nearer goal-post be α (see Figure A).

By considering the right triangles formed by Tina's path, the extension of the goal-mouth, and the lines from Tina to the two goal-posts (see Figure A), we see that

$$\tan \alpha = \frac{x}{y} \quad \text{and so} \quad \alpha = \arctan\left(\frac{x}{y}\right); \tag{1}$$

also

$$\tan(\theta + \alpha) = \frac{x + d}{y} \quad \text{and so} \quad \theta + \alpha = \arctan\left(\frac{x + d}{y}\right). \qquad (2)$$

so

$$\theta = (\theta + \alpha) - \alpha = \arctan\left(\frac{x + d}{y}\right) - \arctan\left(\frac{x}{y}\right)\ldots. \qquad (3)$$

Now for a fixed path (i.e., a fixed value of x) we want to find the value of y that maximizes θ. So, we think of θ as being a function of y, with x and d being constants. Bear in mind, however, that $0 \le x < (w - d)/2$ (see Figure A). Observe that the domain of $\theta(y)$ is $(0, L]$; either $\theta(y)$ will have a global maximum at L, or at an interior point at which $\theta'(y) = 0$. (Clearly $\theta \to \pi/2 - \pi/2 = 0$ as $y \to 0^+$; this is also evident geometrically from Figure A.) Using (3) and the chain rule we differentiate re y and obtain

$$\theta'(y) = \left[1 + \left(\frac{x + d}{y}\right)^2\right]^{-1}\left[\frac{x + d}{y}\right]' - \left[1 + \left(\frac{x}{y}\right)^2\right]^{-1}\left[\frac{x}{y}\right]'$$

$$= \left[1 + \left(\frac{x + d}{y}\right)^2\right]^{-1}\left[-\left(\frac{x + d}{y^2}\right)\right] - \left[1 + \left(\frac{x}{y}\right)^2\right]^{-1}\left[-\frac{x}{y^2}\right].$$

Simplifying, we obtain, after some algebraic manipulation

$$\theta'(y) = -\frac{(x + d)}{y^2 + (x + d)^2} + \frac{x}{(y^2 + x^2)}$$

$$= \frac{d[x(d + x) - y^2]}{[y^2 + (x + d)^2][y^2 + x^2]}.$$

As the denominator of the above fraction is always positive and $d > 0$, $\theta'(y)$ is positive, zero, or negative if and only if its numerator is respectively positive, zero, or negative, i.e., if and only if $x(d + x) - y^2$ is respectively positive, zero, or negative. It quickly follows that $\theta'(y) = 0$ when $y = \sqrt{x(d + x)}$, $\theta'(y) > 0$ when $y \in \left(0, \sqrt{x(d + x)}\right)$, and $\theta'(y) < 0$ when $y \in \left(\sqrt{x(d + x)}, L\right]$. Thus by the First Derivative Test $\theta(y)$ has a global maximum on $(0, L]$ when $y = \sqrt{x(d + x)}$.

However, this assumes that $\sqrt{x(d + x)} \in (0, L]$, so we must verify that this is a reasonable assumption. Clearly $\sqrt{x(d + x)} > 0$, so we must consider whether $\sqrt{x(d + x)} \le L$. From Figure A it is clear that $x \le w/2 - d/2$. Thus $x(d + x) \le (w/2 - d/2)(w/2 + d/2) = (w^2/4) - (d^2/4) < w^2/4$, and so $\sqrt{x(d + x)} < w/2$. Thus $\sqrt{x(d + x)} \le L$ whenever $w/2 \le L$, i.e., whenever $w \le 2L$. Thus so long as the game is played on a "rink" whose width (dimension parallel to the goal-mouths) is no more than twice its length (perpendicular distance between goal-mouths), $\theta(y)$ will have a global maximum at $\sqrt{x(d + x)}$. As every professional hockey rink (and soccer field, etc.) satisfies this constraint, we may assume that the "Salami point" on the path a distance x from the nearer goal-post is a distance $\sqrt{x(d + x)}$ from the extension of the goal-mouth.

By symmetry, it is clear that if we consider a path to the left of the left goal-post that intersects the extension of the goal-mouth at a distance x from the left goal-post, the "Salami point" on that path is again at a distance of $\sqrt{x(d + x)}$ from the goal-mouth extension towards which Tina skates. Finally, if Tina's skating path intersects the goal-mouth, it is clear that the "Salami point" on that path is right on the goal-mouth. Hence if we sketch the portion of the Salami curve to the right of the center of the goal-mouth, the portion to the left of the center of the goal-mouth will be the mirror image of it.

Consider a set of Cartesian coordinate axes with origin at the right-hand goal-post, the extension of the goal-mouth as the X-axis, and the direction toward the opposite goal the positive direction of the Y-axis (see Figure B). The portion of the Salami curve to the right of the right goal-post is the graph of the equation $y = \sqrt{x(d + x)}$, where $0 \le x \le (w - d)/2$. We use standard graph-sketching techniques to sketch this curve. Using the chain rule, we obtain (after some simplification):

$$y'(x) = \frac{2x + d}{2\sqrt{x(d + x)}}.$$

Clearly $y'(x) > 0$ for all x in the domain of y, so this portion of the Salami curve is increasing from left to right. Using the chain rule and the quotient rule, and performing considerable algebraic simplification, we find that

$$y''(x) = \frac{-d^2}{4\sqrt{(dx + x^2)^3}},$$

so clearly $y''(x) < 0$ for all x in the domain of y. Thus this portion of the Salami curve is concave down, and so its shape is roughly as shown in Figure B. Note

The Salami Curve
(to the right of the goal mouth)
one goal mouth width = one unit of distance

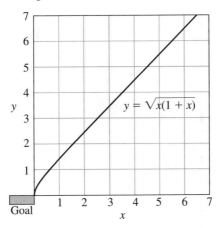

Figure B.

that when x is quite a bit larger than d, then $\sqrt{x(d+x)} \cong x$; hence the portion of the Salami curve far from the goal-mouth lies close to the straight line with equation $y = x$.

Let r denote the perpendicular distance from the goal-mouth to the nearer blue line. Since $y(x)$ increases as x does (as $y'(x) > 0$), we see that the Salami curve meets the blue line if and only if $y((w-d)/2) \geq r$. (To be precise, this uses the continuity of y and the Intermediate Value Theorem.) But

$$y((w-d)/2) = \sqrt{\left(\frac{w-d}{2}\right)\left(d + \left(\frac{w-d}{2}\right)\right)} = \frac{\sqrt{w^2 - d^2}}{2},$$

so the Salami curve meets the blue line if and only if $4r^2 \leq w^2 - d^2$. Once the numerical values of r, w, and d have been obtained for the hockey rink on which you play, you can determine whether the above inequality holds.

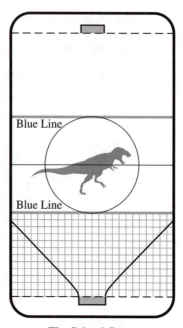

The Salami Curve
The side of a small square = one-fourth
the width of the goal mouth.

Figure C.

Figure C is a moderately accurate scale drawing of the complete Salami curve at one end of the rink. Your icemaker should be able to use it as a blueprint for the real Salami curve.

Respectfully submitted by
Math Iz Us

5

Saving Lunar Station Alpha—Solution

Report No. 3276
Xzyqgon File
Sector: Lunar Station Alpha
Topic: Required launch velocity of LS50 quasion probe rockets for meteor interception

This report addresses the calculation of the launch velocity of the rocket fired from Lunar Station 50 at 3:00 PM on July 21, 21xy, to intercept the meteor on a collision course with Lunar Station Alpha.

Let R denote the radius of Xzyqgon (all distances will be measured in km). Let d denote the distance from Lunar Station 50 (henceforth denoted LS50) to the meteor at 3 PM. Let u denote the speed of the meteor (all speeds in km/sec), and let v denote the maximum possible speed at which the research rockets can be launched.

The situation at 3 PM is shown in Figure A. The circle is the "great circle" on Xzyqgon on which both LS50 and Lunar Station Alpha (henceforth abbreviated LSα) sit. Referring to Figure A, M is the location of the meteor, N is the North Pole where LSα is located, F is the location of LS50, and O is the center of Xzyqgon. Finally, P is the intersection of the path of the research rocket and the path of the meteor, and Q is where the great circle through

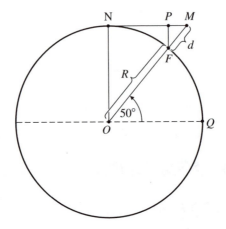

Figure A.

LS50 and LSα meets the Equator. As $OQ \perp ON$ (as the axis of rotation is perpendicular to the plane of the Equator) and $NM \perp ON$ (as the meteor had been calculated to hit LSα tangentially), it is clear that $OQ \parallel NM$.

Now LS50 is on latitude 50° N, so $\angle QOF = 50°$. The rockets at LS50 are pointed 40° north of vertical, so $\angle PFM = 40°$. But $\angle NOF = 90° - 50° = 40°$, so as ON and FP make the same angle (i.e., 40°) with OM, it follows that $ON \parallel FP$. Thus $FP \perp NM$.

The hypotenuse of $\triangle ONM$ is $R + d$; an examination of that triangle yields that $R/(R + d) = \cos 40°$. Solving for d yields

$$d = (R \sec 40°)(1 - \cos 40°)$$

$$= R(\sec 40° - 1). \tag{1}$$

so the value of d can be deduced from the (known) value of the radius R of Xzyqgon.

Let T denote the time (in seconds) that it takes the meteor to travel from M to P—in other words, suppose the meteor reaches P at T seconds after 3 PM. An examination of the right triangle FMP reveals that $MP = d \sin 40°$, and so

$$\frac{d \sin 40°}{u} = T. \tag{2}$$

An examination of $\triangle FPM$ yields that the distance FP from LS50 to the point P where the rocket's path crosses the meteor's path is $d \cos 40°$. Suppose that the rocket is launched at a speed of w km/sec. It will take $(d \cos 40°)/w$ seconds to travel to P, and will arrive at P at $(d \cos 40°)/w$ seconds after 3 PM. So if it arrives at P when the meteor does—i.e., if it hits the meteor—then (using (2))

$$\frac{d\cos 40°}{w} = T = \frac{d\sin 40°}{u}. \qquad (3)$$

Now of necessity $w \leq v$ so from (3) we must have

$$\frac{d\cos 40°}{v} \leq \frac{d\sin 40°}{u}. \qquad (4)$$

if it is possible to hit the meteor with the rocket. Cross-multiplying (4) by the positive quantity $v/(d\sin 40°)$, we see that (4) is equivalent to $u\cot 40° \leq v$; in other words,

$$1.1918u \leq v. \qquad (5)$$

Thus if $1.1918u \leq v$, it is possible to launch the rocket at precisely the speed $1.1918u$ km/sec at precisely 3 PM and have the rocket collide with the meteor, thus assuring the destruction of the meteor.

Now suppose that $1.1918u > v$. Then even if the rocket is launched at its maximum possible speed v, it will not have reached P at T seconds after 3 PM, at which time the meteor will be at P. Suppose that its launch speed is w km/sec. Then T seconds after 3 PM the meteor is at P, and the rocket is between F and P (see Figure B), a distance $d\cos 40° - wT$ km from P.

Now consider the situation t sec after the meteor has passed through P. If the rocket has not yet reached P, then the situation is as portrayed in Figure C. Note that the meteor is ut km from P and the rocket is $d\cos 40° - w(T + t)$ km from P. Let $z(t)$ denote the distance between the meteor and the rocket t sec after the meteor is at P. Applying the Pythagorean theorem to the triangle formed by the rocket, the meteor, and P, we see that

$$z^2 = (d\cos 40° - w(T + t))^2 + (ut)^2. \qquad (6)$$

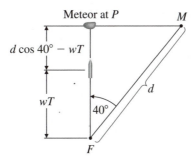

Figure B.

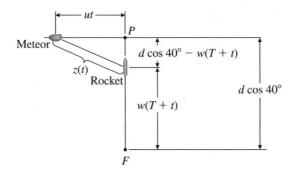

Figure C.

Our objective is to minimize $z(t)$, knowing from our current assumptions that we cannot make $z(t) = 0$, and knowing that $0 \leq w \leq v$.

Observe that before the meteor reaches P, the distance between rocket and meteor is diminishing, and once the rocket is beyond the point P, then the meteor and the rocket are obviously separating (see Figure D). If the speed of the rocket is w, then the time that it takes for the rocket to reach P is clearly $(d \cos 40°)/w$, of which the first T seconds elapse before the meteor reaches P. Hence the relevant domain of z—i.e., the values of t between when the meteor is at P (i.e., at $t = 0$) and when the rocket is at P (i.e., at $t = [(d \cos 40°)/w] - T$)—is the interval $[0, [(d \cos 40°)/w] - T]$. Thus the value that minimizes $z(t)$ lies

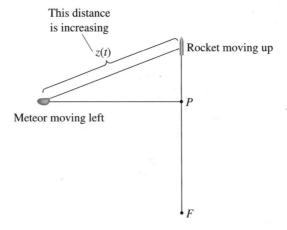

Figure D.

somewhere in that closed interval, either at an end-point or at a critical point in the interior of the interval.

Next we find critical points of $z(t)$. To do so, we differentiate (6) implicitly with respect to t. If we do this we obtain

$$2zz' = 2u^2t + 2[d \cos 40° - w(T + t)][-w].$$

If we simplify, we obtain

$$zz' = u^2t - wd \cos 40° + w^2(T + t). \tag{7}$$

Now $z \neq 0$ (since we are assuming that w is too small for the rocket to hit the meteor) so by (7) we see that $z'(t) = 0$ if and only if

$$u^2t - wd \cos 40° + w^2(T + t) = 0. \tag{8}$$

To classify any critical points we use the second derivative test. Differentiating (7) implicitly re t, we obtain

$$zz'' + (z')^2 = u^2 + w^2. \tag{9}$$

If b were a critical number of z, then $z'(b) = 0$, so substituting $t = b$ into (9) would yield

$$z''(b) = \frac{u^2 + w^2}{z(b)} > 0 \quad \text{(as } z(t) > 0 \text{ for all } t).$$

Thus by the second derivative test $z(t)$ will have a local minimum at any critical point.

Denote $d \cos 40° - wT$ by r. Note that r is the distance that the rocket is from P when the meteor is at P; in particular, $r > 0$. Solving (8) to obtain the value(s) of the critical point(s), we obtain

$$t = \frac{wr}{u^2 + w^2}.$$

Denote this value of t by a. Then $a = (wr)/(u^2 + w^2)$, and (by (6))

$$z(a)^2 = (d \cos 40° - w(T + a))^2 + (ua)^2 = (r - aw)^2 + (ua)^2.$$

Combining these, we obtain

$$z(a)^2 = \frac{u^2w^2r^2}{(u^2 + w^2)^2} + \left(r - \frac{w^2r}{u^2 + w^2}\right)^2,$$

which simplifies to

$$z(a) = \frac{ru}{\sqrt{u^2 + w^2}}. \tag{10}$$

So, the candidates for the minimum value of $z(t)$ on its domain

$$\left[0, \frac{(d \cos 40°)}{w} - T\right]$$

are $z(0)$, $z([(d \cos 40°)/w] - T)$, and $ru/\sqrt{u^2 + w^2}$. A straightforward computation using (6) shows that

$$z(0) = r, \qquad z\left(\frac{d \cos 40°}{w} - T\right) = u\left(\frac{d \cos 40°}{w} - T\right) = \frac{ru}{w},$$

and

$$z(a) = \frac{ru}{\sqrt{u^2 + w^2}}.$$

Since $u < \sqrt{u^2 + w^2}$ and $w < \sqrt{u^2 + w^2}$, it is clear that $z(a)$ is the smallest of the three values and hence represents the minimum distance between rocket and meteor. If we use equations (1) and (2) and the definition of r to write r as a function of R, after some algebraic manipulation we obtain that

$$z(a) = \frac{R(u - w \tan 40°)(1 - \cos 40°)}{\sqrt{u^2 + w^2}}. \qquad (11)$$

Clearly as w increases, the numerator of our expression for $z(a)$ decreases while the denominator increases. Consequently the smallest possible value for $z(a)$ is achieved by launching the rocket at the maximum possible speed, namely v. In this case the smallest separation between rocket and meteor will be

$$\frac{R(u - v \tan 40°)(1 - \cos 40°)}{\sqrt{u^2 + v^2}};$$

when the trig functions are evaluated, this becomes

$$z(a) = \frac{R(u - 0.8391v)(0.234)}{\sqrt{u^2 + v^2}}. \qquad (12)$$

Since the values of R and v are stored in LSα's data banks, and since u has been determined by the use of LSα's Doppler radar, the minimum distance between meteor and rocket could be computed from known data once LSα's computers began to function.

From (3) we see that the launch speed of the rocket that results in the rocket's hitting the meteor is $u \cot 40°$, which equals $1.1918u$. If v is only 98% of this amount—i.e., if $v = (0.98)(1.1918)u = 1.168u$—then by (11) we see that

$$z(a) = \frac{R[u - (0.8391)(1.168u)](0.234)}{\sqrt{u^2 + (1.168u)^2}}$$

$$\cong \frac{(0.02R)(0.234)}{\sqrt{(1 + (1.168)^2}}$$

$$\cong 0.0030436R.$$

If $R = 1000$ km then the closest approach of the rocket to the meteor will be between 3 and 4 km, which is close enough to trigger the proximity fuse, explode the rocket, and deflect the meteor.

Lunar Station Alpha has dodged the bullet!

6

An Income Policy for Mediocria—Solution

To: The Honorable Vanessa Frito
   Minister of Income Policy
   Kingdom of Mediocria
From: Math Iz Us, Inc.
Subject: An income policy for Mediocria

As requested, we present below an analysis of some aspects of an income policy for Mediocria. The basic task is to investigate the way in which income distribution will affect the fraction of their earnings that Mediocrian citizens spend. To this end we establish some preliminary results. In what follows, we shall use data provided by the Mediocrian Census Commission and the Mediocrian Bureau of Statistics. For the moment we denote the relevant quantities by letters.

Let P denote the population of Mediocria. The total amount of money available for the Mediocrian government to pay out in salaries each year is I Mediocrian dollars. Consequently the average annual per capita income A of Mediocrian citizens (measured in Mediocrian dollars) is given by: $A = I/P$.

Denote by $f(x)$ the fraction of the Mediocrian population whose salary is no more than x dollars per year. As P and I are very large, it is an acceptable approximation to assume that the domain of f is the closed interval $[0, I]$; clearly its range is the closed unit interval $[0, 1]$. Let mA denote the largest permissible salary that can be earned in Mediocria (thus m denotes what multiple of the average salary the highest salary is). Clearly $1 \leq m \leq P$, where $m = 1$ if everyone has the same income and $m = P$ if one person receives everything. We wish to demonstrate how P, I, A, m, and f are interrelated. The government can decide both the functional form of f and the size of m, but within the external constraints imposed by the values of I and P. We will further assume that f is a differentiable function; it is easy to imagine situations where this will not be true, but in the highly regimented Mediocrian economy it is a valid assumption.

We begin by showing that m and f are constrained by the requirement that

$$A = \int_0^{mA} xf'(x)\,dx. \tag{1}$$

To prove this, let $\{0 = x_0, x_1, \ldots, x_n = mA\}$ be a partition Q of $[0, mA]$ into n subintervals. We will eventually let $\| Q \|$, the norm of Q, approach zero, so we think of n as being very large, and the length of each subinterval as being very small.

Let $[x_{i-1}, x_i]$ be the ith subinterval of the partition. The fraction of the population earning more than x_{i-1} dollars but not more than x_i dollars annually is

(the fraction earning $\leq x_i$ dollars) $-$ (the fraction earning $\leq x_{i-1}$ dollars)

$= f(x_i) - f(x_{i-1})$.

The total population is P, so the total number of people earning between x_{i-1} and x_i dollars is about $P[f(x_i) - f(x_{i-1})]$.

Denote $x_i - x_{i-1}$ by Δx_i. Then $x_i = x_{i-1} + \Delta x_i$. Applying the Mean Value Theorem to f on $[x_{i-1}, x_i]$, we can find a number $t_i \in [x_{i-1}, x_i]$ such that $[f(x_i) - f(x_{i-1})]/\Delta x_i = f'(t_i)$; hence $f(x_i) - f(x_{i-1}) = f'(t_i)\Delta x_i$. We assume (as noted above) that Δx_i is very small, and so x_{i-1}, x_i, and t_i are nearly equal. So each of the $P[f(x_i) - f(x_{i-1})]$ people earning between x_{i-1} and x_i dollars annually can be assumed to earn t_i dollars annually. Hence all these people together earn about $Pt_i[f(x_i) - f(x_{i-1})]$ dollars annually, i.e., about $Pt_i f'(t_i)\Delta x_i$ dollars annually. If we sum this over all i (i.e., over all income brackets $[x_{i-1}, x_i]$) we obtain approximately the total income received each year by all Mediocrians. In other words,

$$I \cong \sum_{i=1 \text{ to } n} Pt_i f'(t_i)\Delta x_i.$$

In the limit (as $\| Q \| \to 0$) this approximation becomes exact, and we obtain

$$I = \lim_{\| Q \| \to 0} \sum_{i=1 \text{ to } n} Pt_i f'(t_i) \Delta x_i = P \int_0^{mA} x f'(x)\,dx.$$

Now $I/P = A$, so dividing the above by P, we obtain

$$A = \int_0^{mA} x f'(x)\,dx, \tag{1}$$

and so equation (1) is verified as required.

We can integrate (1) by parts, using the formula $\int u\,dv = uv - \int v\,du$, with $x = u$ and $f'(x)\,dx = dv$. This gives

$$\int_0^{mA} x f'(x)\,dx = x f(x) \Big|_0^{mA} - \int_0^{mA} f(x)\,dx$$

$$= mA f(mA) - (0)f(0) - \int_0^{mA} f(x)\,dx.$$

Now $f(mA)$ is the fraction of the population that receives income less than or equal to the largest income received by anyone, so evidently $f(mA) = 1$. Hence equation (1) becomes

$$A = mA - \int_0^{mA} f(x)\,dx.$$

The difference D between the maximum income mA and the average income A is therefore given by

$$D = mA - A = \int_0^{mA} f(x)\,dx.$$

We next show that if $r(x)$ is the fraction of income spent by a Mediocrian whose annual income is x dollars per year (Mediocrians are remarkably uniform in their spending habits!), then the total amount S spent by all Mediocrians in a year is given by

$$S = \int_0^{mA} P x r(x) f'(x)\,dx. \tag{2}$$

Again, because of the large size of I and P, it is reasonable to think of the domain of r as being $[0, I]$; clearly its range is a subset of $[0, 1]$. Small changes in income do not provoke drastic changes in spending habits among Mediocrians, so we may assume that r is a continuous function.

Instead of presenting a formal proof of (2) using partitions, as we did for (1), we give a more informal argument using the language of infinitesimals of the sort that a physical scientist or engineer might use. We trust that you can change it into the language of partitions if need be.

Let dx be an "infinitely small positive number." The fraction of the population earning between x and $x + dx$ dollars per year is (as we argued when establishing (1)) $f(x + dx) - f(x)$. The total number of such people is $P[f(x + dx) - f(x)]$. Each earns about x dollars per year, and therefore saves $xr(x)$ dollars per year. Thus the total amount saved each year by all Mediocrians in this income bracket is $xr(x)P[f(x + dx) - f(x)]$ dollars. Multiplying and dividing by the positive quantity dx, and realizing that dx is "very small," we see that this equals

$$xr(x)P[f(x + dx) - f(x)] = xr(x)P \left[\frac{f(x + dx) - f(x)}{dx} \right] dx$$

$$\cong xr(x)Pf'(x)\, dx.$$

So dS, the amount saved annually by Mediocrians in this (infinitely narrow) income bracket, is given by

$$dS = Pxr(x)f'(x)\, dx.$$

Thus the total amount S saved by all Mediocrians each year is given by

$$S = \int dS = \int_0^{mA} Pxr(x)f'(x)\, dx,$$

which is as claimed in equation (2). (Again we assume that f is differentiable.)

We have been requested to construct a model of income distribution in which all levels of income, from 0 to mA, are represented by the same number of people. As seen earlier, the number of people with incomes between x and $x + dx$ is approximately $f'(x)\, dx$, so our model requires that $f'(x)$ be a constant (say K). Consequently we must have $f(x) = \int K\, dx = Kx + C$, where C is another constant.

Let α be the fraction of the population that has no income at all. Then $\alpha = f(0) = C$, so $f(x) = Kx + \alpha$. As noted before, $f(mA) = 1$, so $1 = KmA + \alpha$, which yields $K = (1 - \alpha)/mA$. Thus

$$f(x) = \left[\frac{1 - \alpha}{mA} \right] x + \alpha, \tag{3}$$

and

$$f'(x) = \frac{1 - \alpha}{mA}.$$

Substituting this into equation (1) yields

$$A = \int_0^{mA} x \left[\frac{1-\alpha}{mA} \right] dx = \left[\frac{1-\alpha}{mA} \right] \left[\frac{x^2}{2} \right] \Big|_0^{mA} = \left[\frac{1-\alpha}{2} \right] [mA].$$

As $A \neq 0$ (if $A = 0$, then $I = 0$ and Mediocria is in deep trouble), it follows that $1 = [(1-\alpha)/2]m$, and so $m = 2/(1-\alpha)$. As $0 \leq \alpha < 1$ (since if $\alpha = 1$ then $I = 0$), it follows that $0 < 1 - \alpha \leq 1$ and so $m \geq 2$. Thus in order for this economic policy to be implemented, the maximum income will have to be at least twice the average income.

Also observe that since $m = 2/(1-\alpha)$, it follows that $1 - \alpha = 2/m$, and so

$$f(x) = \left(\frac{2}{m^2 A} \right) x + \left(1 - \frac{2}{m} \right), \tag{4}$$

and

$$f'(x) = \frac{2}{m^2 A}. \tag{5}$$

We now calculate the total annual spending of Mediocrian citizens under the assumptions that we have been given. Clearly it will depend on the value of m. We will denote the total annual spending by Mediocrians by $S(m)$ if the maximum allowable income is mA. We are also assuming that (4) and (5) hold, and that

$$r(x) = 1 - \left(\frac{x}{4A} \right) \quad \text{if } 0 \leq x \leq 3A;$$
$$r(x) = \tfrac{1}{4} \quad\quad\quad \text{if } 3A \leq x.$$

Thus the range of r is $[0, 1]$ and r is continuous as specified. As $m \geq 2$ and the formula specifying $r(x)$ changes when $x = 3A$, we consider two cases separately.

Case 1 $2 \leq m \leq 3$.

Then by equation (2) we have

$$S(m) = \int_0^{mA} P x r(x) f'(x) \, dx$$

$$= P \int_0^{mA} x \left[1 - \left(\frac{x}{4A} \right) \right] \left[\frac{2}{m^2 A} \right] dx$$

$$= \left[\frac{2P}{m^2 A} \right] \int_0^{mA} \left[x - \left(\frac{x^2}{4A} \right) \right] dx$$

$$= \left[\frac{2P}{m^2A}\right]\left[\frac{6Ax^2 - x^3}{12A}\Big|_0^{mA}\right]$$

$$= PA\left[1 - \left(\frac{m}{6}\right)\right].$$

But $PA = I$, so $S(m) = I[1 - (m/6)]$. Thus the fraction $F(m)$ of total income that is spent annually by Mediocrians is given by

$$F(m) = 1 - (m/6) \quad \text{if } 2 \leq m \leq 3.$$

This ranges in value from $2/3$ (when $m = 2$) to $1/2$ (when $m = 3$). (We remark that the conspicuous lack of attractive consumer goods may be the cause of such a low spending rate.)

Case 2 $m \geq 3$.

In this case, again using equation (2), we have

$$S(m) = \left[\int_0^{3A} Pxr(x)f'(x)\,dx\right] + \left[\int_{3A}^{mA} Pxr(x)f'(x)\,dx\right]$$

$$= \left[P\int_0^{3A} x\left[1 - \left(\frac{x}{4A}\right)\right]\left[\frac{2}{m^2A}\right]dx\right] + \left[\int_{3A}^{mA}(Px)\left(\frac{1}{4}\right)\left(\frac{2}{m^2A}\right)dx\right]$$

The first term in the above sum was evaluated in Case 1 with an upper limit of mA rather than $3A$. If we use that calculation, we see that the first term in the above sum equals

$$\left[\frac{2P}{m^2A}\right]\left[\frac{6Ax^2 - x^3}{12A}\right]\Big|_0^{3A} = \frac{9PA}{2m^2} = \frac{9I}{2m^2}.$$

Now

$$\int_{3A}^{mA}(Px)\left(\frac{1}{4}\right)\left(\frac{2}{m^2A}\right)dx = \left(\frac{P}{2m^2A}\right)\left[\int_{3A}^{mA} x\,dx\right]$$

$$= \left(\frac{P}{2m^2A}\right)\left(\frac{A^2(m^2 - 9)}{2}\right)$$

$$= \frac{I}{4} - \frac{9I}{4m^2}.$$

Adding the two terms, we see that

$$S(m) = I\left[\frac{1}{4} + \frac{9}{4m^2}\right],$$

and so

$$F(m) = \frac{1}{4} + \frac{9}{4m^2}.$$

Thus $F(m)$ decreases as m increases, but no matter how wealthy Mediocria is, with this income policy and these spending habits, the fraction of income that Mediocrians spend will always exceed $1/4$.

Respectfully submitted,

Math Iz Us

7

The Case of the Cooling Cadaver—Solution

To: Inspector Vance McFrito
From: Math Iz Us
Subject: Calculation of the time of death of Lord Boddy

We present below our estimation of the time of death of Lord Boddy, who passed away sometime on the night of June 23–24, 20xy, as the result of a severe blow to the back of the head administered with a heavy brass candlestick. Since only one suspect failed to have an alibi at the time of death, we can therefore conclude who committed the murder.

We accuse Professor Prune of murdering Lord Boddy in the study with the candlestick at approximately 11:28 PM on June 23, 20xy. The reasoning that led us to this conclusion follows.

Let T denote the temperature of the study. (All temperatures will be measured in degrees Celsius.) According to evidence presented by Sherlock Marples, and because of the precautions taken to keep the study unoccupied throughout the investigation, we may assume that T remained constant throughout the period of the investigation.

Let t denote the number of hours elapsed from the time of the murder. Let a denote the number of hours from the time of the murder until 1:30 AM. Let

$C(t)$ denote the temperature of the corpse t hours after the time of the murder. Let $y(t)$ denote the difference in temperature between corpse and study t hours after the murder. Then

$$y(t) = C(t) - T.$$

Since the study is rather cool, clearly $y(t) \geq 0$ throughout our investigation.

One version of Newton's Law of Cooling states that the rate of change of the temperature of a cooling body is directly proportional to the difference between the body and its surroundings (if the surroundings remain at a constant temperature). In symbols:

$$C'(t) = ky(t),$$

where k is a constant depending on the physical properties of the cooling body (its size, shape, specific heat, etc.). Now

$$y'(t) = \frac{d}{dt}[C(t) - T] = C'(t) - \frac{dT}{dt} = C'(t) - 0 = C'(t)$$

since, as T is a constant, $dT/dt = 0$. Hence $y'(t) = ky(t)$, which, as we know, has as its solution $y(t) = y(0)\exp(kt)$. Thus $C(t) - T = y(0)\exp(kt)$, i.e.,

$$C(t) = T + y(0)\exp(kt). \tag{1}$$

This gives the temperature of the corpse t hours after the murder.

Lord Boddy's body temperature (when he was alive) was the usual 37° Celsius, so we know that $C(0) = 37$. Substituting $t = 0$ into (1), we obtain

$$37 = C(0) = T + y(0)\exp(0) = T + y(0).$$

Thus $y(0) = 37 - T$ and hence (1) becomes

$$C(t) = T + (37 - T)\exp(kt). \tag{2}$$

At 1:30 AM (i.e., when $t = a$) we know that the temperature of the corpse was 32° Celsius. Thus

$$C(a) = 32. \tag{3}$$

At 2:30 AM we have $t = a + 1$ and $C(t) = 30$, so

$$C(a + 1) = 30. \tag{4}$$

At 3:30 AM we have $t = a + 2$ and $C(t) = 28.25$, so

$$C(a + 2) = 28.25. \tag{5}$$

Combining each of (3), (4), and (5) with (2) yields the following equations:

$$32 = T + (37 - T)\exp(ka); \tag{6}$$

$$30 = T + (37 - T)\exp(k(a + 1)); \tag{7}$$

$$28.25 = T + (37 - T)\exp(k(a + 2)). \tag{8}$$

Solve (6) for exp(*ka*) and get

$$\frac{32 - T}{37 - T} = \exp(ka). \tag{9}$$

Using this, we can rewrite (7) as

$$\frac{30 - T}{37 - T} = \exp(ka + k) = \exp(ka)\exp(k)$$

$$= \left(\frac{32 - T}{37 - T}\right)\exp(k).$$

Solve this for exp(*k*) and get

$$\frac{30 - T}{32 - T} = \exp(k). \tag{10}$$

Using (9) and (10), we can now rewrite (8) as

$$28.25 = T + (37 - T)(\exp(ka))(\exp(2k))$$

$$= T + (37 - T)\left(\frac{32 - T}{37 - T}\right)\left(\frac{30 - T}{32 - T}\right)^2,$$

and so

$$28.25 = T + \frac{(30 - T)^2}{(32 - T)}.$$

Thus

$$(28.25 - T)(32 - T) = (30 - T)^2.$$

When this is expanded, the terms in T^2 cancel, leaving a linear equation in T whose solution is easily computed to be: $T = 16$. Substituting this into (10) gives

$$\exp(k) = \frac{30 - 16}{32 - 16} = \frac{7}{8}.$$

Substituting this into (9) gives

$$\frac{32 - 16}{37 - 16} = (\exp k)^a = \left(\frac{7}{8}\right)^a.$$

Thus $\ln(16/21) = \ln((7/8)^a) = a\ln(7/8)$. Hence

$$a = \frac{[\ln((16/21)]}{[\ln(7/8)]} \cong 2.036.$$

Now $.036 \times 60 \cong 2$, so $a \cong 2$ hours and 2 minutes. Thus 1:30 AM is about 2 hours and 2 minutes after the time of the murder, placing the time of the murder at about 11:28 PM. As Professor Prune is the only suspect without an alibi at that time, he clearly must be the murderer.

8

Designing Dipsticks—Solution

To: Joe Moosemess
From: Anne G. Gables
Subject: Dipstick Calibration

Here is my report on how to calibrate the dipsticks for the cylindrical and spherical fuel tanks. In each case I have worked out an algebraic expression giving the fraction $F(d)$ of the tank capacity that is filled with fuel as a function of the length d of dipstick that is wetted by the fuel. I have also provided a table of values for $F(d)$ for values of d ranging from 0 to D (the value of d when the tank is full) in increments of $D/32$. For the cylindrical tank I have derived the algebraic expression in two ways. (You will be reassured to note that they agree!)

Cylindrical Tanks

Consider a right circular cylindrical fuel tank whose circular ends are vertical. Let the radius of its ends be R, and let its length be h (all distances will be measured in meters). Suppose that a dipstick inserted in the slot at the top is wetted to a length d. This means that the surface of the fuel in the tank intersects the circular ends of the tank along a line that is a distance d above the bottom of

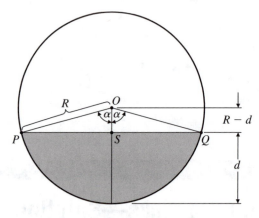

Figure A.

the end of the tank (see Figure A). Observe that d will equal $2R$ when the tank is full, and in general $0 \le d \le 2R$; Figure A is drawn for the case $0 \le d \le R$. Let $A(d)$ denote the area of the circular tank end that will be wetted by the fuel. Obviously the volume $V(d)$ of fuel in the tank is given by $hA(d)$.

Let the center of the circular end of the tank be denoted by O, and let P and Q denote the points where the "wetting line" from the fuel surface intersects the outside of the circular tank end (see Figure A). We calculate $A(d)$ in two ways. First we will use some geometry (but no calculus) to calculate $A(d)$ in terms of d and R in the case where $0 \le d \le R$. (Once this is done we show that the same formula will hold if $R \le d \le 2R$). Let S be the point where a perpendicular through O intersects PQ; as $OP = OQ$, S is the midpoint of PQ. Let $\angle QOS = \alpha$ (all angles are measured in radians). It is clear from Figure A that

$$A(d) = (\text{area of sector } POQ) - (\text{area of } \triangle POQ). \qquad (1)$$

Now sector POQ subtends an angle of 2α radians, so using the standard formula for the area of a sector of a circle, we see that

$$\text{area of sector } POQ = \frac{2\alpha}{2\pi}(\pi R^2) = \alpha R^2.$$

Using $\triangle QOS$ we see that $\cos \alpha = (R - d)/R$, and so $\alpha = \arccos((R - d)/R)$. Hence

$$\text{area of sector } POQ = R^2 \arccos\left(\frac{R - d}{R}\right).$$

Applying the Pythagorean theorem to $\triangle QOS$, we see that $QS = \sqrt{2Rd - d^2}$. Using the formula for the area of a triangle, one sees that

$$\text{area of } \triangle POQ = (1/2)(2)\sqrt{2Rd - d^2}(R - d).$$

Putting these values into equation (1) gives

$$A(d) = R^2 \arccos((R - d)/R)$$
$$- \sqrt{2Rd - d^2}(R - d) \quad (0 \leq d \leq R).\dots \quad (2)$$

If $R \leq d \leq 2R$ then it is evident that

$$A(d) = \pi R^2 - A(2R - d).$$

Now $0 \leq 2R - d \leq R$ since $R \leq d \leq 2R$, so we can use (2) to evaluate $A(2R - d)$. If we do this, after some algebraic simplification we obtain that

$$A(d) = R^2[\pi - \arccos(-(R - d)/R)] - \sqrt{2Rd - d^2}(R - d).\dots \quad (3)$$

It is a straightforward exercise in trigonometry to show that $\pi - \arccos(-s) = \arccos(s)$ if $0 \leq s \leq 1$, and so it follows from (3) that

$$A(d) = R^2 \arccos((R - d)/R)$$
$$- \sqrt{2Rd - d^2}(R - d) \quad (R \leq d \leq 2R).\dots \quad (4)$$

Combining (2) and (4) gives

$$A(d) = R^2 \arccos((R - d)/R)$$
$$- \sqrt{2Rd - d^2}(R - d) \quad (0 \leq d \leq 2R).\dots \quad (5)$$

Now we use calculus to calculate $A(d)$. Consider a circle of radius R placed on a set of coordinate axes with center at $(R, 0)$ (see Figure B). Its equation is $y^2 + (x - R)^2 = R^2$; the portion above the X-axis has equation $y = \sqrt{R^2 - (x - R)^2}$. Now think of this circle as being one end of the cylindrical tank, with the "up" direction being in the direction of the positive X-axis. If $0 \leq d \leq 2R$, we see that

$$A(d) = 2 \int_0^d \sqrt{R^2 - (x - R)^2} \, dx.$$

We integrate this by using the trig substitution $x = R + R \cos \theta$. Then $dx = -R \sin \theta \, d\theta$ and $\sqrt{R^2 - (x - R)^2} = R \sin \theta$. (Observe that $0 \leq x \leq d \leq 2R$; hence $0 \leq \theta \leq \pi$, and so $R \sin \theta \geq 0$.) When $x = 0$, then $\cos \theta = -1$, and so

$\theta = \pi$. When $x = d$, then $\cos\theta = (d - R)/R$, and so $\theta = \arccos((d - R)/R)$. Thus our integral becomes

$$A(d) = 2 \int_{\pi}^{\arccos((d-R)/R)} (-R^2 \sin^2\theta)\, d\theta$$

$$= 2 \int_{\arccos((d-R)/R)}^{\pi} (R^2 \sin^2\theta)\, d\theta.$$

It is routine to show that $\int \sin^2\theta\, d\theta = (\theta/2) - ((\sin 2\theta)/4) = (\theta/2) - (\sin\theta\cos\theta)/2$. When $\theta = \arccos((d - R)/R)$ we have $\cos\theta = (d - R)/R$ and $\sin\theta = \sqrt{2Rd - d^2}/R$, so evaluating at the limits of integration gives

$$A(d) = 2R^2 \left[\frac{\pi}{2} - \frac{1}{2}\arccos\left(\frac{d - R}{R}\right) + \frac{(d - R)\sqrt{2Rd - d^2}}{2R^2} \right]$$

$$= R^2 \left[\arccos\left(\frac{R - d}{R}\right) + (d - R)\frac{\sqrt{2Rd - d^2}}{R^2} \right]$$

$$= R^2 \arccos\left(\frac{R - d}{R}\right) - \sqrt{2Rd - d^2}(R - d) \qquad (0 \le d \le 2R),$$

which is the same value as in equation (5).

Now the volume of a cylindrical tank of radius R and length h is $\pi R^2 h$ so the fraction $Fc(d)$ of the capacity of the tank that has fuel in it when the dipstick wets to a length d is given by

$$Fc(d) = \frac{hA(d)}{\pi R^2 h} = \frac{\arccos\left(\frac{R-d}{R}\right) - \sqrt{2Rd - d^2}\left(\frac{R-d}{R^2}\right)}{\pi}. \qquad (6)$$

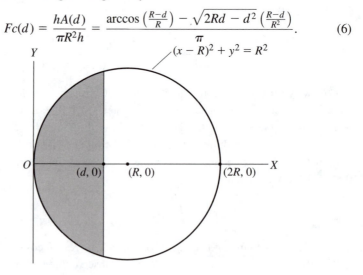

Figure B.

Using a scientific calculator (I would have used a computer algebra system had one been available at the lodge), I calculated values for $Fc(d)$, as d ranges from 0 to $2R$ in integer multiples of $R/16$. The results are appended in Table 1. [AUTHOR'S NOTE: This has not been included.]

Spherical Tanks

A spherical tank of radius R can be generated by rotating about the X-axis a circle with radius R and center at $(R, 0)$ (see Figure B). If we think of the "top" of the tank as being at $(2R, 0)$, then the volume $V(d)$ of fuel in the tank when the dipstick (lying along the X-axis) wets to a length d is just the volume of the solid generated by rotating about the X-axis the region bounded below by the X-axis, above by a portion of the circle, and to the right by the line $x = d\,(0 \le d \le 2R)$. As seen above, the portion of the circle above the X-axis has equation $y = \sqrt{R^2 - (x - R)^2} = \sqrt{2Rx - x^2}$. Hence, using the standard formula for the volume of a solid of revolution, we see that

$$
\begin{aligned}
V(d) &= \int_0^d \pi y^2 \, dx \\
&= \int_0^d \pi[2Rx - x^2] \, dx \\
&= \pi R x^2 - \frac{\pi x^3}{3} \Big|_0^d \\
&= \frac{\pi d^2 (3R - d)}{3}.
\end{aligned}
$$

The volume (fuel capacity) of the tank is $(4\pi R^3)/3$, so the fraction $Fs(d)$ of the capacity of the tank that has fuel in it when the dipstick wets to a length d is given by

$$
Fs(d) = \frac{\dfrac{\pi d^2 (3R - d)}{3}}{\dfrac{4\pi R^3}{3}},
$$

and so

$$
Fs(d) = \frac{d^2 (3R - d)}{4R^3}. \tag{7}
$$

I calculated values for $Fs(d)$, as d ranges from 0 to $2R$ in integer multiples of $R/16$. The results are appended in Table 2. [AUTHOR'S NOTE: This has not been included.]

Let $d = 2mR$, where $0 \leq m \leq 1$; then m is the fraction of the length of the dipstick that is wetted when it is inserted in the tank. Substituting this value of d into (6) and (7) and simplifying, we obtain

$$Fc(2mR) = \frac{1}{\pi} \left[(\arccos(1 - 2m)) - 2(1 - 2m)\sqrt{m - m^2} \right]. \tag{8}$$

and

$$Fs(2mR) = 3m^2 - 2m^3. \tag{9}$$

In each case the expression for the fraction of fuel in the tank is independent of the size of the tank.

✤ ✤ ✤ ✤ ✤

Since Billy used the spherical dipstick and concluded that the cylindrical tank was 0.232 full, the spherical dipstick was wetted to a length that was calibrated as 0.232. To find the fraction of the dipstick that was wetted, we must solve the equation $0.232 = 3m^2 - 2m^3$ (see equation (9)). If we do this (using a computer algebra system, Newton's method, or some other device), we find that $m \cong 0.3125$. Substituting this into equation (8), we obtain

$$Fc((2)(0.3125)(1)) \cong 0.267.$$

So the cylindrical tank is actually about 0.267 full, rather than about 0.232 full.

9

The Case of the Gilded
Goose-Egg—Solution

Part 1: Spherical Eggs

To: Members of the Pirate Crew
From: Della Friday
Concerning: Equal division of a gold-plated spherical "goose-egg"

Your captain has asked me to show you how to slice your gold-plated spherical "goose-egg" in such a way that each crew member receives the same amount of gold. He has asked me to denote both the radius of the egg and the number of crew members by letters, in case division of a spherical egg of a different radius among a different number of crew members should become an issue at some time in the future. In what follows, distances can be measured in any convenient units; I will not specify any.

A sphere of radius R is generated by rotating a circle of radius R (with center at the origin) about the X-axis.

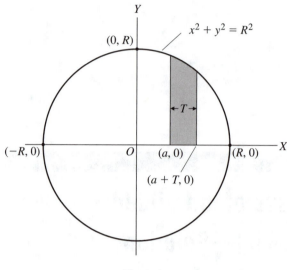

Figure A.

A "slice" of such a sphere of thickness T is similarly generated by rotating about the X-axis a region of the circle bounded between vertical lines with horizontal separation T, above by a portion of the circle, and below by the X-axis (see Figure A).

Our circle will have equation $x^2 + y^2 = R^2$, so the semicircle above the X-axis has equation $y = \sqrt{R^2 - x^2}$. Hence a strip of this circle of width T (which, when rotated about the X-axis, produces a slice of thickness T) will be bounded between the line $y = 0$, the vertical lines $x = a$, and $x = a + T$ (where $-R \le a < a + T \le R$), and the circle $y = \sqrt{R^2 - x^2}$. Let d denote the thickness of the gold-plating of the sphere. As d is very small compared to R, the volume of gold on the above-mentioned slice is essentially equal to (surface area of slice)(d).

Using the standard formula for the surface area of a surface of revolution, and denoting the outside surface area of our slice by S (measured in square units), we see that

$$S = \int_{a}^{a+T} 2\pi y \sqrt{1 + (y')^2}\, dy. \qquad (1)$$

Here $y = \sqrt{R^2 - x^2}$,

so

$$y' = \left(\frac{1}{2}\right)\left(\frac{1}{\sqrt{R^2 - x^2}}\right)(-2x) = \frac{-x}{\sqrt{R^2 - x^2}}.$$

Thus

$$2\pi y \sqrt{1 + (y')^2} = 2\pi\sqrt{R^2 - x^2}\sqrt{1 + \left(\frac{-x}{\sqrt{R^2 - x^2}}\right)^2}$$

$$= \left(2\pi\sqrt{R^2 - x^2}\right)\sqrt{\frac{R^2}{R^2 - x^2}}$$

$$= 2\pi R.$$

Substituting this into (1), we see that the volume V of gold on the outside of this slice, which (as noted above) equals Sd, is given by

$$V = d\int_a^{a+T} 2\pi R\, dx$$

$$= 2\pi R\, d(a + T) - 2\pi R\, da$$

$$= 2\pi R\, dT.$$

Thus the volume of gold-plating on this slice of the sphere is independent of a, i.e., is independent of the distance of the slice from one end of the sphere. It depends only on the thickness T of the slice. Thus if we set T to be $2R/n$ and then slice the sphere into n slices of equal thickness, each person will obtain the same amount of gold, i.e., $2\pi R\, d(2R/n) = 4\pi R^2\, d/n$ cubic units of gold.

Part 2: Elliptical Eggs

For the eyes of Captain P. King only

My dear Captain,

As we determined from our measurements on board your ship, your "goose-egg" is in fact an ellipsoid of revolution, obtained by rotating an ellipse about the longer of its two axes. If an ellipse is placed on a coordinate axis system so that its center is at the origin and its long axis lies along the X-axis then its equation will be

$$\left(\frac{x}{a}\right)^2 + \left(\frac{y}{b}\right)^2 = 1, \tag{2}$$

where its semi-major axis and semi-minor axis are a and b respectively. Thus $0 < b \leq a$. If we solve (2) for the positive value of y, we obtain

$$y = \frac{b}{a}\sqrt{a^2 - x^2}, \tag{3}$$

and this will be the equation of the portion of the ellipse above the X-axis.

A "slice of goose-egg" of thickness T is obtained by rotating about the X-axis a region bounded between the vertical lines $x = s$, $x = s + T$, bounded below by the X-axis, and bounded above by the curve with equation $y = (b/a)\sqrt{a^2 - x^2}$ (see (3)), where the parameter s measures from where in the ellipse the strip to be rotated is chosen (see Figure B). As both the lines $x = s$ and $x = s + T$ must intersect the ellipse, it is necessary that $-a \leq s < s + T \leq a$; in other words, it is necessary that $s \in [-a, a - T]$. Since the thickness of the gold-plating (which we again denote by d) is very small in comparison to a and b, the volume $V(s)$ of gold on the outside of the slice is essentially dS, where S denotes the surface area of the portion of the slice from the outside of the ellipsoid. (As this volume may depend on the position of the strip within the ellipse, which is measured by s, we denote the volume as being a function of s.) Using equation (1) from our analysis of the sphere in Part 1, we see that

$$V(s) = d \int_{s}^{s+T} 2\pi y \sqrt{1 + (y')^2}\, dy. \tag{4}$$

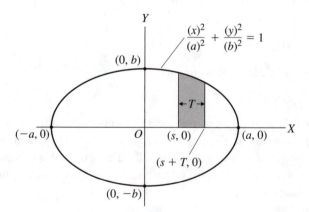

Figure B.

Let us denote b/a by r; then $0 < r \leq 1$. Thus (3) becomes

$$y = r\sqrt{a^2 - x^2},$$

and so

$$y' = \frac{-rx}{\sqrt{a^2 - x^2}}.$$

A routine computation then gives

$$\sqrt{1 + (y')^2} = \sqrt{\frac{a^2 - (1 - r^2)x^2}{a^2 - x^2}}$$

from which we obtain

$$2\pi y \sqrt{1 + (y')^2} = 2\pi r \sqrt{\frac{[a^2 - x^2][a^2 - (1 - r^2)x^2]}{a^2 - x^2}}$$

$$= 2\pi r \sqrt{a^2 - (1 - r^2)x^2}.$$

Combining this with equation (4) above we see that

$$V(s) = 2\pi r d \int_s^{s+T} \sqrt{a^2 - (1 - r^2)x^2}\, dx, \qquad (5)$$

and as noted above, $V(s)$ (regarded as a function of s) has domain $[-a, a - T]$. We want to find at what point(s) on the interval $[-a, a - T]$ $V(s)$ assumes its maximum value. To do this we evaluate $V(s)$ at $-a$, at $a - T$, and at any critical points (i.e., points c for which $V'(c) = 0$). To this end we next compute $V'(s)$.

If g is a continuous function with domain $[p, q]$ then a version of the Fundamental Theorem of Calculus tells us that if $G(s)$ is defined to be $\int_p^s g(x)\, dx$ for each $s \in [p, q]$ then $G'(s) = g(s)$ for each $s \in [p, q]$ (using one-sided derivatives at the endpoints). Hence, using the Chain Rule,

$$\frac{d}{ds}(G(s + T)) = G'(s + T) \cdot \frac{d}{ds}(s + T) = g(s + T) \cdot 1$$

$$= g(s + T) \quad (\text{if } p \leq s < s + T \leq q).$$

Consequently

$$\frac{d}{ds}\left[\int_s^{s+T} g(x)\, dx\right] = \frac{d}{ds}\left[\int_p^{s+T} g(x)\, dx - \int_p^s g(x)\, dx\right]$$

$$= \frac{d}{ds}\left[\int_p^{s+T} g(x)\, dx\right] - d/ds\left[\int_p^s g(x)\, dx\right]$$

$$= g(s + T) - g(s).$$

If we apply this to equation (5), with $\sqrt{a^2 - (1 - r^2)x^2}$ playing the role of $g(x)$, $-a$ playing the role of p, and $a - T$ playing the role of q, we obtain

$$V'(s) = 2\pi r d \sqrt{a^2 - (1 - r^2)(s + T)^2}$$
$$- 2\pi r d \sqrt{a^2 - (1 - r^2)s^2} \quad \text{(for } -a \leq s \leq a - T\text{)}. \quad (6)$$

If we set $V'(s) = 0$ and simplify (remembering that $r > 0$ and $d > 0$), we find that $V'(s) = 0$ iff $(1 - r^2)(s + T)^2 = (1 - r^2)s^2$. Now $r = 1$ iff $a = b$, in which case our ellipsoid is a sphere, $V'(s)$ is identically zero, and we are back to the earlier case. If we have a true ellipsoid, then $r < 1$, so $1 - r^2 > 0$ and we see that $V'(s) = 0$ iff $(s + T)^2 = s^2$. As $T > 0$, this has one root, namely $s = -T/2$. As the thickness of a slice cannot be more than $2a$ (the length of the ellipsoid), we have that $0 < T \leq 2a$ and so $-a \leq -T/2 < 0 \leq a - T$. Hence our critical point is in the domain of $V(s)$. We will classify it by using the First Derivative Test.

As $-T/2$ is the only point of $[-a, a - T]$ at which $V'(s) = 0$, by the Intermediate Value Theorem $V'(s)$ does not change sign on $[-a, -T/2)$ or on $(-T/2, a - T]$. As the number n of people among whom the goose-egg must be divided is certainly bigger than 2, and as each slice of egg is to have the same thickness, it follows that $T \leq 2a/3$, and so $-T \in [-a, -T/2)$. Using equation (6), we calculate that

$$V'(-T) = 2\pi r d [a - \sqrt{a^2 - (1 - r^2)T^2}].$$

Since $0 < 1 - r^2 < 1$ and $T < a$, we see that $[a^2 - (1 - r^2)T^2]^{0.5} < a$ and so $V'(-T) > 0$. Thus $V'(s) > 0$ if $s \in [-a, -T/2)$. A similar calculation (whose details will be omitted) yields that $V'(0) < 0$, and so $V'(s) < 0$ if $s \in (-T/2, a]$. Thus $V(-a) < V(-T/2)$ and $V(a - T) < V(-T/2)$, so $V(s)$ assumes its maximum value on $[-a, a - T]$ at $-T/2$. To compute that maximum value we would have to evaluate the integral in equation (5). This is possible, but messy, and since it was not specified in our contract, we have not done so.

We next must determine what numbers will occur as X-coordinates of left-hand edges of strips that are rotated to form slices, and under what conditions $-T/2$ is one of those numbers. As the goose-egg is to be divided among n crew members and each slice is to have the same thickness T, it is evident that we must have $T = 2a/n$. The left-hand edge of the jth slice from the left must therefore have equation $x = -a + (j - 1)T = -a + (j - 1)(2a/n)$ (where j is some integer from 1 to n). As $-T/2 = -a/n$, we need to find an integer j for which $-a + (j - 1)(2a/n) = -a/n$. When we solve this for j, we obtain

$j = (n + 1)/2$. If n is an odd integer, say $n = 2m + 1$, then $j = m + 1$, and the $(m + 1)$st slice from the left, which is formed by rotating the strip bounded between $x = -a/(2m + 1)$ and $x = a/(2m + 1)$, is the slice with the most gold. If n is an even integer, then the two slices adjacent to the center, which are swept out by the strips bounded by $x = -2a/n$ and $x = 0$, and by $x = 0$ and $x = 2a/n$, are congruent and contain the same amount of gold, which is more than any other slice contains. (Captain King, I will allow you to verify that for yourself!)

In closing, you always want to take the slice as close to the center of the ellipsoid as you can if you wish to maximize your gold holdings.

Respectfully submitted,
Della Friday
(Della Friday, B.Sc., LL.B.)

10

Sunken Treasure—Solution

To: Captain P. King
From: Della and Joseph Friday
Subject: Sinking the treasure barge

First of all, we want to tell you that we are very disappointed by your conduct. Even pirate kings ought to know the basic rules of etiquette, and one of the most fundamental is that well-bred hosts do not knock out their guests and kidnap them.

However, enough of that. You have asked us to calculate the exact length needed of a roll of material in which a barge with a parabolic cross-section will be slung and lowered to the ocean floor. By measuring certain dimensions of the barge, we can calculate the shape of the parabola. When the barge is at rest on the ocean floor it will sit upright—i.e., the vertex of its parabolic cross-section will sit on the ocean floor so that the axis of symmetry of the parabolic cross-section will be vertical. We know the vertical distance from the ocean floor to the tops of the four pylons—call this distance b. (All distances will be measured in meters.) We know the horizontal perpendicular distance from axle to boom—we will denote it by $2a$. We also know that the sunken barge will be centered exactly between the boom and the axle, so that the perpendicular distance from the boom to the axis of symmetry of the parabola,

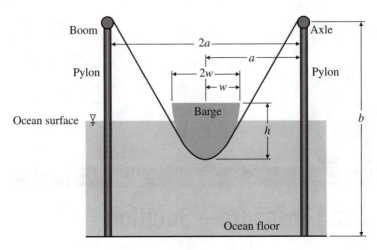

Figure A.

and from the axle to the axis of symmetry of the parabola, is a. The width of the deck of the barge will be denoted by $2w$, and the depth of the barge (the perpendicular distance from the deck to the bottom of the hull of the barge at the vertex of the parabola) will be denoted by h. This information is summarized in Figure A. In Figure B we impose a set of coordinates with the origin at the vertex of the parabola, the X-axis the trace of the ocean floor on our diagram, and the Y-axis the axis of symmetry of the parabola. With respect to this system of coordinates the equation of the parabolic cross-section is $y = kx^2$, where $k > 0$. It is clear that the point (w, h) is on the parabola, so $h = kw^2$, whence $k = h/w^2$. Although we do not know what the exact relationships among our parameters will be in general, it is clear that if the barge is to fit between the pylons, then $w < a$. As the barge is to be completely submerged, we must have $h < b$. Of course, all distance parameters are positive numbers.

When the barge sits on the ocean floor and the roll of material forming a sling for the barge is pulled tight, the trace on Figure B of that roll consists of a section of the parabolic trace of the hull together with a line segment from the hull to the trace of the boom in Figure B (i.e., the point R). As there is no slack in the sling, there are two possibilities, depending on the relative sizes of the parameters a, b, h, and w. First, the sling might meet the hull along a line whose trace on Figure B is a point Q whose X-coordinate (call it c) is less than w. Second, the sling might enclose the entire hull and then change direction abruptly from the top of the hull to the axle (see Figure C). We consider these cases in turn.

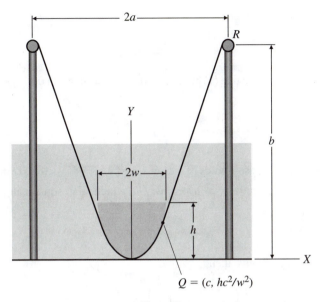

$$Q = (c, hc^2/w^2)$$

Figure B.

Case 1   The trace of the sling in Figure B meets the hull at a point Q whose X-coordinate c is less than w. In this case, because the sling is pulled tight, part of the trace of the sling forms a line segment RQ that is tangent to the parabola at Q. Let c be the X-coordinate of Q; then Q has coordinates (c, kc^2). From the dimensions specified earlier we see that R has coordinates (a, b). (We remark that there are two tangents to the parabola that pass through R, but from geometric considerations we want the one for which $0 < c < w < a$.)

We first compute c in terms of these quantities by equating two expressions for the slope of the line segment RQ. Using the "rise over run" definition, we see that

$$\text{slope of } RQ = \frac{b - kc^2}{a - c}.$$

As RQ is tangent to the parabola at Q, we see that

$$\text{slope of } RQ = y'(c) = 2kc.$$

Thus $\dfrac{b - kc^2}{a - c} = 2kc$, which simplifies to $kc^2 - 2kac + b = 0$. Solving for c using the quadratic formula yields

$$c = a \pm \sqrt{a^2 - \frac{b}{k}} = a \pm \sqrt{a^2 - \frac{bw^2}{h}}.$$

These roots are real iff

$$\frac{bw^2}{h} \leq a^2. \tag{1}$$

Assume for the moment that (1) holds. As we are assuming that $0 < c < w < a$, we choose the root

$$c = a - \sqrt{a^2 - \frac{bw^2}{h}}. \tag{2}$$

Thus assuming (1), Case 1 holds iff $c \leq w$, i.e., if $\sqrt{a^2 - \frac{bw^2}{h}} \leq w$. Transposing and squaring yields $(a - w)^2 \leq a^2 - (bw^2/h)$, which, after some manipulation, yields

$$w \leq (2ah)/(b + h). \tag{3}$$

Observe that if (3) holds, then

$$\frac{bw^2}{h} \leq \left(\frac{4a^2h^2}{(b+h)^2}\right)\left(\frac{b}{h}\right) = a^2\left(\frac{4bh}{(b+h)^2}\right). \tag{4}$$

But $(b + h)^2 - 4bh = (b - h)^2 \geq 0$ so $4bh/(b + h)^2 \leq 1$. Thus (1) follows from (4), and hence from (3). Hence Case 1 holds iff (3) holds. Thus once the parameters a, b, w, and h are measured, one can check quickly whether the Case 1 hypotheses hold.

From Figure B it is clear that in Case 1 the total length L of the roll of material that we will use for the sling is given by

$$L = 2(L_1 + L_2), \tag{5}$$

where

$$L_1 = \text{the distance from } R \text{ to } Q$$

and

$$L_2 = \text{the distance along the parabola from the origin } O \text{ to } Q.$$

We want to express L in terms of the known quantities a, b, h, and w.

Using the distance formula, we see that

$$L_1 = \sqrt{(b - kc^2)^2 + (a - c)^2}$$

$$= \sqrt{\left(b - h\left(\frac{c}{w}\right)^2\right)^2 + (a - c)^2}. \tag{6}$$

Using the formula for arc length along a curve, we see that

$$L_2 = \int_0^c \sqrt{1 + y'(x)^2}\,dx$$

$$= \int_0^c \sqrt{1 + 4k^2 x^2}\,dx.$$

Let $2kx = \tan\alpha$. Then $2k\,dx = \sec^2\alpha\,d\alpha$ and $\sqrt{1 + 4k^2 x^2} = \sqrt{1 + \tan^2\alpha} = \sec\alpha$. When $x = 0$, then $\alpha = 0$, and when $x = c$, then $\alpha = \arctan(2kc)$. Hence, after some manipulation, we see that

$$L_2 = \frac{1}{2k} \int_0^{\arctan(2kc)} \sec^3\alpha\,d\alpha.$$

When this standard integral is evaluated and the limits of integration and the value of k are substituted, we obtain

$$L_2 = \left(\frac{c}{2}\right)\left(\sqrt{1 + \frac{4h^2 c^2}{w^4}}\right) + \ln\left[\frac{2hc}{w^2} + \sqrt{1 + \frac{4h^2 c^2}{w^4}}\right]. \qquad (7)$$

Substituting (6) and (7) into (5), and (2) into the result, yields an expression for L in terms of a, b, h, and w. It is too messy to write out explicitly.

Case 2   If $c > w$ (which by (3) above will happen when $w > (2ah)/(b+h)$), then we have the situation illustrated in Figure C; the sling wraps around the hull up to the point $(w, h) = T$, and its trace on Figure C is possibly non-differentiable at T. In this case it is evident that the length L of sling needed is given by $L = 2(L_1 + L_2)$, where (by reasoning similar to that used in Case 1)

$$L_1 = \sqrt{(b - kw^2)^2 + (a - w)^2}$$

$$= \sqrt{(b - h)^2 + (a - w)^2},$$

and

$$L_2 = \frac{1}{2k} \int_0^{\arctan(2kw)} \sec^3\alpha\,d\alpha;$$

i.e.,

$$L_2 = \left(\frac{w^2}{4h}\right)\left(\frac{2hw}{w^2}\right)\sqrt{1 + \frac{4h^2 w^2}{w^4}}$$

$$+ \ln\left[\frac{2hw}{w^2} + \sqrt{1 + \frac{4h^2 w^2}{w^4}}\right]$$

$$= \left(\frac{w}{2}\right)\sqrt{1 + \frac{4h^2}{w^2}} + \ln\left[\frac{2h}{w} + \sqrt{1 + \frac{4h^2}{w^2}}\right].$$

As in Case 1, substituting these expressions for L_1 and L_2 into $L = 2(L_1 + L_2)$ gives L in terms of the measured parameters a, b, h, and w.

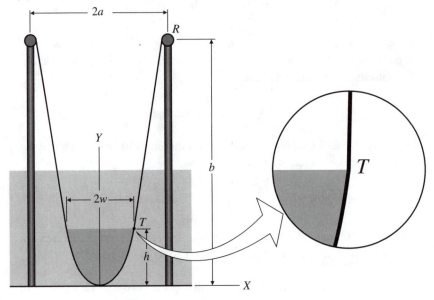

Figure C.

❖ ❖ ❖ ❖ ❖

We have taken careful measurements of the barge and the boom-and-axle arrangement for the particular barge that you want to sink now. The deck is 30 m wide, so $w = 15$. The other measured values were: $h = 27$, $a = 40$, and $b = 70 + 14 = 84$ (all distances in meters). Then $(2ah)/(b + h) \cong 19.4$ while $w = 15$. Hence inequality (3) holds, and we are in a "Case 1" situation. Substitution into (2) gives $c = 10$, and further substitution into (6) gives $L_1 = 78$. Substitution into (7) then gives $L_2 = (5)(2.6) + (1/0.48)(\ln 5) \cong 16.35$. Thus in the present case we have $L \cong 2(78 + 16.35) = 188.7$. Thus you will have almost exactly 188.7 m of material unrolled from the axle when the barge comes to a gentle rest on the bottom of the ocean.

Prepared by:

Della Friday

Joe Friday

11

The Case of the Alien Agent—Solution

TOP SECRET

For the eyes of the Core Cell of Alienwatch only.

Analysis of the freutcaquium layer

As you may know, one of your members anonymously sent us top secret documents giving the calculation by the IPP of the volume of freutcaquium fallout in the Earth's crust. He (or she) pointed out that this calculation did not take into account the curvature of the Earth, and might therefore have overestimated the actual amount of freutcaquium that settled on our planet. If this were true, the amount of freutcaquium might have been insufficient to wipe out the ancient aliens that you (and we) believe may have populated the Earth at that time.

However, our calculations seem to indicate that the error made by the IPP did not cause their results to be incorrect by more than at most 3%, which is probably within the experimental error in the field measurements obtained by the IPP geologists. We present our calculations below, and invite you to verify them.

❖ ❖ ❖ ❖ ❖

The layer of freutcaquium can be thought of as a solid of revolution obtained as follows. Consider a circle K of radius R with center at the origin. Above the portion of the circle in the first quadrant consider a curve C with the following property: if $0 \le \theta \le \alpha \le \pi/4$, and if $L(\theta)$ is a line through the origin O lying in the first quadrant and making an angle θ with the Y-axis, then the distance along $L(\theta)$ from O to C is $R + s(\theta)$, where $s(\theta)$ is defined to be $d - (d\theta/\alpha)$ (see Figure A). Thus C and K meet when $\theta = \alpha$, and this marks the end of C. The resulting diagram is the picture obtained by slicing through the Earth with a plane; the origin O is the center of the Earth, the point $(0, R)$ is the point of impact of the ancient asteroid, and C is the trace on the plane of the top of the layer of freutcaquium. Hence R represents the radius of the Earth (all distances are in meters), and d represents the thickness of the layer of freutcaquium at the point of asteroid impact. Evidently $d \ll R$, and we will use

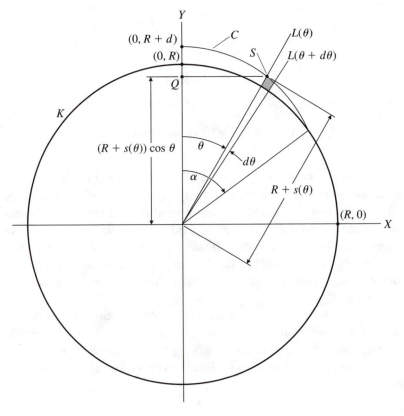

Figure A.

this fact to make several simplifying approximations. If the region J bounded between the circle, the Y-axis and C is rotated about the Y-axis, the resulting solid of revolution represents the total freutcaquium layer. We must calculate this volume.

For a momentarily fixed value of θ, consider the infinitesimal portion $P(\theta)$ of J bounded by C, K, and the lines $L(\theta)$ and $L(\theta + d\theta)$. Its width varies from $R\,d\theta$ at K to $(R + s(\theta))\,d\theta$ at C. As $0 \leq s(\theta) \leq d \ll R$ for all $\theta \in [0, \alpha]$, we can approximate the width to be $R\,d\theta$ throughout. If $P(\theta)$ is rotated about the Y-axis, it sweeps out a shell of thickness $R\,d\theta$ that forms part of the surface area of a right circular cone obtained by rotating the right triangle OQS about the Y-axis (see Figure A). A line segment in the first quadrant of length a and inclination θ ($0 \leq \theta \leq \alpha$), with one end at the origin, has equation $y = x \tan \theta$ ($0 \leq x \leq \alpha \cos \theta$); see Figure B. Using the formula for the surface area of a surface of rotation, we see that this line segment, when rotated about the X-axis, sweeps out a surface of area

$$
\begin{aligned}
S &= \int_0^{a \cos \theta} 2\pi y \sqrt{1 + (y')^2}\, dx \\
&= \int_0^{a \cos \theta} 2\pi \tan \theta \sqrt{1 + \tan^2 \theta}\, x\, dx \\
&= (2\pi \tan \theta \sec \theta)\frac{x^2}{2}\bigg|_0^{a \cos \theta} \\
&= \pi a^2 \sin \theta.
\end{aligned}
$$

Using this result, we quickly see (denoting $s(\theta)$ by s) that the surface area of the shell swept out by rotating the infinitesimal region $P(\theta)$ about the Y-axis is

$$
\pi(R + s)^2 \sin \theta - \pi R^2 \sin \theta = (\pi s \sin \theta)(2R + s).
$$

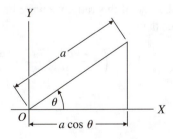

Figure B.

As $s \ll R$ as noted earlier, we can approximate this as $2\pi Rs \sin\theta$. Hence the volume dV of ash generated by rotating $P(\theta)$ about the Y-axis is given by

$$dV = (2\pi Rs \sin\theta)(R\,d\theta).$$

Thus the total volume V of freutcaquium deposited is given by

$$V = \int_0^\alpha 2\pi R^2 \left(d - \frac{d\theta}{\alpha} \right) \sin\theta\,d\theta$$

$$= 2\pi R^2 d \int_0^\alpha \left(\sin\theta - \frac{1}{\alpha}\theta \sin\theta \right) d\theta.$$

A routine integration by parts shows that

$$\int_0^\alpha \frac{1}{\alpha}\theta \sin\theta\,d\theta = \frac{\sin\alpha - \alpha\cos\alpha}{\alpha},$$

and of course $\int_0^\alpha \sin\theta\,d\theta = (-\cos\alpha) + 1$. Consequently

$$V = 2\pi R^2 d \left[-\cos\alpha + 1 - \frac{1}{\alpha}\sin\alpha + \cos\alpha \right];$$

i.e., $\quad V = \left(2\pi R^2 \frac{d}{\alpha} \right)(\alpha - \sin\alpha).$ \hfill (1)

The IPP computed the volume V^* of freutcaquium by assuming that the Earth was flat, in which case its shape would be a right circular cone with altitude d and base radius r, where r is the (great circle) distance from the impact crater to where the thickness of the freutcaquium layer has decreased to zero. Evidently $V^* = \pi r^2 d/3$ and $r = R\alpha$, so $V^* = \pi R^2 \alpha^2 d/3$. Thus, comparing the sizes, we see that

$$V/V^* = \frac{6}{\alpha^3}(\alpha - \sin\alpha). \hfill (2)$$

To get an estimate for the size of this ratio, we write out the Maclaurin series for $\sin\alpha$; using it, we see that

$$\frac{6}{\alpha^3}(\alpha - \sin\alpha) = \frac{6}{\alpha^3} \left[\alpha - \left(\alpha - \frac{\alpha^3}{3!} + \frac{\alpha^5}{5!} - \frac{\alpha^7}{7!} - + \cdots \right) \right]$$

$$= \frac{6}{\alpha^3} \left[\frac{\alpha^3}{3!} - \frac{\alpha^5}{5!} + \frac{\alpha^7}{7!} - + \cdots \right]$$

$$= 1 - \frac{\alpha^2}{20} + \frac{\alpha^4}{840} - + \cdots$$

Now the magnitude of the difference between the infinite sum and the sum of the first n terms of a convergent alternating series in which the magnitudes of successive terms is decreasing is no larger than the magnitude of the $(n + 1)$st term. As $0 \le \alpha \le \pi/4 < 1$, the magnitude of successive terms in this series is indeed decreasing and we can conclude that

$$1 - \frac{\alpha^2}{20} + \frac{\alpha^4}{840} < \frac{V^*}{V} < 1.$$

Now $\alpha \le \pi/4$ and so $1 - \alpha^2/20 + \alpha^4/840 > 0.97369$, as a quick calculator computation verifies. Consequently $V > V^* > 0.97V$, and the amount of freutcaquium calculated by the IPP is over 97% of the "real" amount of freutcaquium; in fact, the approximation is probably considerably better than that, depending on the size of α. We reluctantly conclude that the results sent by the IPP to the TFARB, even if formally calculated incorrectly, are probably very close to the correct value. However, we feel sure that there must be a government conspiracy involved in this work somewhere. Keep looking! The truth is out there.

Muldie and Sculler.